D1785056

G.T.I.S. LIBRARY
THE BRITISH ALUMINIUM CO. LTD.
CHALFONT TECHNOLOGICAL CENTRE,
CHALFONT PARK,
GERRARDS CROSS, BUCKS.

8 - 8 - 74

ENERGY
IN
THE
1980s

ENERGY
IN
THE
1980s

A ROYAL SOCIETY
DISCUSSION ORGANIZED BY
SIR PETER KENT, F.R.S.

LONDON
THE ROYAL SOCIETY
1974

ISBN 0 85403 062 X

This report was first published in
Philosophical Transactions of the Royal Society
of London, series A, volume 276, no. 1261,
pages 405–615

Printed in Great Britain for the Royal Society
at the University Printing House, Cambridge

Preface

By Sir Peter Kent, F.R.S.
Natural Environment Research Council, London, WC1H 0AX

The proposal for a discussion meeting on 'Energy in the 1980s' was made by the Royal Society Industrial Activities Committee early in 1972. It was felt that concern about the technical availability of different forms of fossil fuel justified an authoritative review; that this should be linked with consideration of alternative energy sources and with the problems of energy conversion and power generation. The efficient utilization of energy was also recognized as critical, but as requiring separate and deliberate consideration on another occasion.

The detailed programme was planned by a committee consisting of Mr George Armstrong of the National Coal Board, Professor J. S. Forrest, F.R.S., of the Central Electricity Generating Board, Mr G. P. Glass of Shell U.K. Limited, Dr W. C. Marshall, F.R.S., of the United Kingdom Atomic Energy Authority, and the writer. In line with the general trend of the Royal Society industrial review symposia the range was limited to energy sources likely to be of importance in the next decade, omitting subjects which were more directly relevant to the 1990s and the next century. This involved a degree of judgement in selection which may prove to have not been entirely accurate; under the new sharply increased pressures for alternative energy requirements it now seems that solar power, in particular, might become important as an auxiliary power source in some countries rather sooner than was envisaged only a year ago.

The Royal Society was fortunate in obtaining the services of the Secretary of State for Trade and Industry, Rt Hon. Peter Walker, M.P., to give the opening address, and is particularly grateful to him for carrying out his undertaking to speak despite the coincidence of the meeting with the height of the energy crisis and the potential difficulty of discussing national policy at such a time.

The opening address was followed by two papers on statistical studies – the first by S. H. Schurr & J. Darmstadter of Resources for the Future Inc., of Washington D.C., which gave an optimistic view of energy prospects (particularly in the fossil fuel range), and the second by D. C. Ion of the World Energy Conference, who provided a most timely reminder of the widely differing ways in which energy statistics were derived and used.

An account of world coal resources was provided by D. J. Ezra, Chairman of the National Coal Board; in his absence, due to sickness, the paper was read by G. Armstrong, who called attention to a major downward adjustment of coal reserve statistics related partly to the necessity for increasingly sophisticated extraction methods. Sir Eric Drake, Chairman of the British Petroleum Company Limited, dealt with oil reserves, stressing that the world discovery rate of major fields had fallen progressively since the middle 1960s despite greatly improved methods of prospecting, and concluding that in about five years time an increasing gap between oil demand and supply could be expected to develop. C. P. Coppack of Shell International Gas Ltd, demonstrated the misfit between the distribution of natural gas reserves and world markets, showing that regional shortages already developing were liable to become an increasingly dominant factor.

[v]

In an account of world hydro energy K. R. Vernon of the North of Scotland Hydro-Electric Board demonstrated the advanced stage of development of hydro power in the industrialized countries (stressing the limitation on complete water utilization dictated by the necessity for maintaining adequate river flow for amenities, industrial use, removal of sewage, etc.) in contrast to the very low degree of utilization in the developing countries. During discussion he pointed out that extremely high capital cost eliminated tidal energy, and also wind power, as practicable large scale energy sources.

S. H. U. Bowie of the Institute of Geological Sciences described current and probable future supplies of natural nuclear fuels, an energy source contrasting with the previous cases in being sharply price-dependent, with a potential for greatly increased production given high prices. Professor T. Leardini of Ente Nationale per L'Energia Electrica, Rome, spoke on geothermal energy, its sporadic distribution world-wide, and on the need for much more widespread investigation of its potentialities.

The second day was largely concerned with the broader principles of energy conversion, beginning with a discussion of coal utilization by L. Grainger of the National Coal Board, followed by an account of the potential and current state of development of non-conventional hydrocarbon sources (oil shale and tar sands) by E. B. Walker of Gulf Oil Corporation and a discussion on primarily engineering lines of gas transmission problems by D. E. Rooke of British Gas Corporation. The increasing importance of nuclear sources, combined with energy storage systems, was stressed by D. Clarke of the Central Electricity Generating Board; Dr T. Broom, also from the Central Electricity Generating Board gave an encouraging account of operational experience in British nuclear power stations, and Sir John Hill of U.K.A.E.A. completed the technical presentations by discussing future trends in nuclear power generation.

We were indebted to Dr G. B. R. Feilden of British Standards Institution for undertaking the onerous task of summarizing the proceedings of the meeting and the critical problem of the future balance of energy reserves. The meeting coincided with a world-wide fuel crisis which underlined the immediacy of this problem, and the requirement for planning an energy future largely different from that of the early 1970s.

The Royal Society, and the organizing committee of 'Energy in the 1980s', are deeply grateful to the authors who have made possible this symposium, as well as to those who contributed to the very lively discussions at the meeting.

The members of the organizing committee would also like to include in this acknowledgement Miss Patricia Ritchie of the Royal Society staff, who was responsible for most of the administrative work and for the smooth running of the conference.

CONTENTS

CONTENTS

FUTURE BALANCE OF ENERGY RESOURCES

Phil. Trans. R. Soc. Lond. A. **276**, 407–412 (1974)
Printed in Great Britain

Opening address

By The Rt Hon. Peter Walker, M.P.
Secretary of State for Trade and Industry

In a week in which we are threatened with an immediate crisis in coal and an immediate crisis in electrical power and a threatening shortage of oil, I welcome the opportunity to open this two-day meeting that the Royal Society has organized. I want to speak particularly of the practical task confronting the country over the next 20 years. Before doing this I should like to make a few general observations.

In deference to my audience I will first mention some figures which illustrate the magnitudes we are dealing with. Total world consumption of energy in 1961 was estimated to be some 1.2×10^{20} J (4.2×10^9 tonnes of coal equivalent). By 1970 this had grown by 60 % to 1.9×10^{20} J (6.7×10^9 t c.e.). By 1985 exponential projections suggest the figure may be 4.5×10^{20} J (nearly 16×10^9 t c.e.) – $2\frac{1}{2}$ times more in 15 years.

These numbers show the massive scale of what we are considering. But I must now reveal my own view about such projections. Virtually all energy predictions of both demand and supply that have been made in the past have proved to be wrong. I believe in spite of all of our experiences and lessons that have been learnt from inaccurate predictions in the past it is almost equally difficult to predict the total world-wide trends in the future. There are so many uncertainties. On the supply side new energy resources, as yet unknown, can always be discovered. Even 10 years ago few would have predicted the massive resources that we have now discovered in the seas surrounding our coasts. Few can predict accurately the full implications of the development of the fast breeder reactor and the transformation it could bring to the potentialities of nuclear energy. Massive research is taking place throughout the world into new forms of energy and major breakthroughs could be achieved. As to the consumption of energy, the development of the electric motor vehicle could transform the pattern of consumption in one obvious sphere. Substantial savings could be made by far more positive policies towards the elimination of wastage of energy. All this does not of course mean that we should not make forecasts. We must make the best ones we can, recognizing that they may be wrong. Where they suggest a shortage of energy we have to take vigorous action to prove them wrong. One difficulty when making forecasts is to calculate the massive increase in energy consumption that will take place as a result of the new wealth that is currently going to the primary producing countries of the world. The huge increase in commodity prices is going to give a whole range of world economies, from Brazil to Nigeria, from Iran to Venezuela to some of the countries of Africa, the finest opportunity to raise living standards, to raise the speed of policies for their industrialization and therefore to raise enormously their energy consumption.

When a year ago I took on my present responsibilities, I naturally reviewed the then current developments in energy policy. The outstanding point of course was that the British economy as with the economies of the European Community and of Japan had a heavy dependence upon imported oil: a dependence which had been increased in the whole of the post-War period, increased as a result of the fact that throughout the world energy from oil was cheap and abundant. In fact it fitted into the natural objective of governments that their energy policy should be based upon meeting their energy requirements with reasonable security and lowest total costs over time. But, in the changing scene that I inherited of increasing uncertainty as to both price and supply, it was vital that I pursued a policy to maximize the economic use of all indigenous resources and to do my best to ensure that improved supplies were obtained at the most reasonable price and with the most reasonable security that was available. In this situation therefore a number of major decisions were taken and a practical policy developed.

In the short term, one has to build up defences against temporary interruptions of supply. This meant developing an improved stock position and also having contingency plans which could be applied if there was a necessity to share out a shortage in the least harmful manner. Many months before the present crisis policies were pursued to achieve this objective.

When our oil supplies were threatened with disruption I was able to announce that we had 79 days of stocks with a further 30 days stocks on the way to this country. I might say that those stocks have been an expensive charge upon our balance of payments but, as events have proved, a sensible insurance premium to pay. Likewise, I took powers under the Coal Industry Act to meet the cost of building up substantial stocks of coal. We have built up stocks at power stations of no less than 20×10^6 tonnes – equivalent to 12 weeks supply.

The next facet of our positive approach to energy policy was to examine all of the indigenous resources that were available to us and see that they were developed as swiftly and effectively as possible. I think the nation has underestimated the transformation of energy policy that has taken place over the past 12 months.

There was the decision to reverse previous policies and to stop the rapid rundown of the coal industry. Between 1964 and 1970, 274 pits had been closed, 185 000 mining jobs had been lost and coal production had dropped by 50×10^6 t. I decided in November last year that looking at all of the potential sources of energy supply that would be available to Britiain in the 1980s coal would be an important ingredient. I therefore reversed the policy. I introduced the Coal Industry Act which contributed no less than £1100M to achieve this purpose. It meant that jobs would be maintained, that financial help would be given to the selling of coal, that conditions with regard to pensions and redundancy benefits would be improved and that the financial position of the National Coal Board would be transformed.

Since that time I have been engaged in reviewing with the National Coal Board their future investment programmes. Our problem is to develop the viable core, solve the problems of the peripheral fields and continue to make substantial gains in productivity. I must, however, give a warning as to the immediate contributions that the coal industry can make. The fact is that over the next 10 years obviously a number of existing collieries will run out of coal. The building of a new colliery is, alas, a very long-term project. For example, we have discovered a very important seam of coal near Selby in Yorkshire – a seam that could well be the finest seam of coal ever discovered in the United Kingdom. But if all of the current tests prove positive and we make available all of the financial resources that are needed to develop that seam, such are the technical tasks involved that it would in fact be 8–10 years before that new pit was in full production.

Next we have the tremendous potential of oil in the North Sea. It was only a few months ago that I was being criticized for pursuing a policy that would speedily extract the maximum volume of oil from this source. The events of the last two months have seemingly silenced my critics. On all of our calculations of energy problems of the next 20 years it must be right to bring the maximum acceleration to the task of extracting oil from this particular source. There is of course considerable public misunderstanding as to the timing: a misunderstanding that in days when I have been less engaged with the energy problem I would probably have shared. There is a feeling that the oil has been discovered and so the view to quote a letter from a lady who wrote in recently to the B.B.C. programme 'Any answers': 'Why do we bother to take oil all the way from the Middle East, why not bring back our tankers from the Middle East and fill them up from the North Sea'. Alas, it is a complicated and expensive task. It is only a few years ago that these resources were discovered. Already £1000M has been invested in the development of these resources and as yet no oil has been landed and there will be very little before 1975.

I would express the hope that with the massive natural interest in this topic and the rather fascinating technical and physical problems involved some enterprising television corporation will soon make an effective film, explaining the difficulties of extracting oil from this Sea. What in fact has perhaps been amazing is the speed of deep-sea petroleum exploration and extraction. Six years ago the deepest water production of the world was around 60 m. North Sea oil will be coming from places where the sea bed is over 120 m deep. There are exciting developments such as semi-submersible drilling platforms which will operate in all but the worst weathers, underwater well completions and reinforced-concrete oil platforms, but with all of these designs there are problems which will need to be overcome in the deeper waters and there is little doubt that the eventual need will be for a completely underwater production system which will raise fresh problems of design and working conditions.

The other sphere of North Sea development, of immense importance is the situation as far as the extraction of gas is concerned. It really is a remarkable achievement of both those that have extracted the gas and of the gas industry in this country that already 90 % of our gas supplies come from the North Sea. It was absolutely right of the Government to pursue the policy of maximizing the resources available to Britain by negotiating with our friends in Norway so that, subject to the ratification of the Norwegian Parliament, we will be able to extract the gas from the part of the Frigg Field that is under their influence. There are possibilities of purchasing gas from elsewhere, but they are less attractive, much more expensive, and some such sources of course have all of the problems of the political stability that we have in terms of oil extraction in other parts of the world.

Both in the oil and gas in the North Sea it must be the policy of the Government to bring these developments to fruition as rapidly as possible. It was for this purpose that we made contact with all of the oil companies. We have endeavoured to find out the difficulties, the problems and the bottlenecks with which they are confronted – for example, steel supplies. We have seen that they have had priority of allocation because of the great national importance of their task. They have had problems of labour relations which we hope will improve, particularly with the recognition of the massive national importance of bringing these schemes to fruition. And certainly it is the intention of my Department to keep in close contact with all of those developing the North Sea, to see that any action that Government can take to assist them in speeding up these developments will in fact be made.

Having dealt with the way in which we have endeavoured to improve the situation in coal, in gas and in oil – I will turn, if I may, to nuclear energy. This is really an indigenous form of energy for all practical purposes, and it will increasingly replace fossil fuels as they become expensive and scarce. And ultimately perhaps a very considerable proportion of the energy of the world will be supplied by this particular means. It was for this reason that in the review of energy policy that I carried out just a year ago I decided to bring about a very substantial reorganization of the whole of our nuclear industry. I decided it was vital to expand out nuclear power base as quickly as was practical. As a first step it was important to strengthen the nuclear industry by encouraging the creation of one unified design and construction firm. We now have the National Nuclear Corporation. I then wanted to see that the major decisions, such as choice of reactors, were taken on the best possible advice obtainable. It was for this reason that I set up the Nuclear Power Advisory Board with myself as the Chairman of that Board, and contained within that Board the leading authorities and expertise that is available in this country. I have immediately given that Board the task of looking at the immediate problems of nuclear policy, both national and international, and of course we do have an immediate task of choosing a type of reactor. No decision of any description has been taken yet, but the attempt, obviously, is to choose for the interim stages of nuclear development the type of reactor which will best assist us in the longer term. There is no doubt at all in our view, and this is a shared view I think by all concerned, that in the longer term the fast-breeder reactor has the considerable advantage of a better use of uranium and is seen as the predominant type. The prototype development is going to be critical next year and we would expect to be able to place the first commercial order for the fast-breeder reactor in 1976, with a full programme of fast-breeder reactors coming into development and orders being placed throughout the 1980s.

Once again the time-lag is very considerable, and there is no possibility of making decisions today on nuclear policy which will result in any major contribution to our energy supplies within the crucial few years from now until the end of this decade. But in the decade thereafter there is no doubt in my mind that nuclear power will be a very fast-emerging source of energy. International cooperation is important on nuclear development because of the huge costs – the need, for example, to provide fuel for increased nuclear programmes – and governments are seeking this. We have entered into cooperation with Germany and Holland on producing plant for centrifuge-enrichment of uranium. It is vital that we keep in close contact with our colleagues in Europe and colleagues in other parts of the world who are involved in the development of nuclear energy in all of its forms. The new company is doing this and, of course, my own Advisory Board is keeping in close contact with it. And, therefore, in this sphere there are important developments that have taken place in the reorganization. There are very important decisions to be made shortly, and over the next two or three years we are, I think, going to see very considerable progress with our nuclear policy.

Therefore, summing up the energy policy at home, it is very simply to develop as quickly and swiftly as possible the maximum utilization of the indigenous resources that are available to us. Already the major decisions have been taken to do this in coal, gas, oil and in nuclear energy. But there is also, of course, the need, and there will continue to be the need, to import energy from abroad – certainly over the next decade. These imports are vital to us, and it is important to establish the best relationships we can with the producers of these energy supplies. And that is why I am very anxious to see as far as the oil-producing countries of the Middle East are concerned, given the stages of development that some of those countries will now go through in processes of industrialization and in activities diversifying their economies, that Britain plays a very close role with those nations and helps and assists in every possible way. I believe that it is important for the British economy, which has this unique advantage of the considerable resources of the North Sea available to it in the 1980s, to realize that perhaps it has a strong and important negotiating position in studying the long-term economic and commercial interests –

[4]

not just of this country but of the countries that are energy producers at the present time. And I would hope that from this, better commercial and economic relationships could result, and the result of these could be greater security of our oil supplies in this country.

There are, of course, other important developments. The potentialities of the oil shales and the tar sands are enormous. But once again there is a very considerable time-lag involved. Although the increased price of energy has made all of these potentially much more interesting and possible, once you have made the decision that they are to make an important contribution, the time-lag between that decision and actually getting large supplies of energy from those immense resources quite a considerable one and not an immediate solution to our problems.

There is perhaps a more immediate solution and help in better utilization of the energy resources that we enjoy. And here I think it is important that governments and industry should take action during a period of considerable world energy problems to try and reduce the considerable wastage that does take place in our energy resources. Quite a great deal can be achieved in the future by proper insulation of buildings, and governments can act – and I believe will act – in this in terms of the building regulations they produce. But, of course, the immediate impact is on new buildings, whereas most of the energy is used in old buildings. And although one might be able to do further actions as far as old buildings are concerned once again there is a considerable time-scale involved. There is no doubt at all that industry can obtain and agree on a much better industrial use of energy. And certainly by studies that have already been made by a number of individual industries, there is considerable potentiality for reducing present levels of energy usage.

The electric car is a potentiality; although of course still a user of energy, is it of a different nature and different scale, and will give more flexibility to the whole system. Here there are considerable problems primarily concerned with the battery involved. The government has taken an active interest in this and has already provided grants to see what is possible in further research and development of the sodium–sulphur battery.

To sum up, at the moment this country obtains 50 % of its energy from overseas and 50 % is indigenous. We would anticipate from the forecasts that are available that by 1980 we could be providing 75 % of our energy from indigenous sources and that figure can rise thereafter beyond that point, depending upon what in total is found in the resources of the North Sea, and thereafter on a longer scale depending upon the success of our nuclear developments. It does place Britain in a better position as far as energy is concerned than almost all of our major industrial competitors. It does present us with opportunities of developing new industries that will help the world cope with their future energy problems. I believe, for example, the expertise that we will develop in the offshore drilling of the North Sea will give a whole range of British firms an expertise and know-how which will have very considerable world-wide application in the years ahead.

Likewise, I believe the developments of the fast-breeder reactor and developments in the sphere of nuclear energy will have an important world-wide application of benefit to the British economy, and British commerce as a whole. And so although we are faced with immediate considerable problems of energy in the domestic scene, and although we are faced with a considerable dependence upon imported energy for the coming five to ten years, we do have opportunities which are quite unique and do present this country with a considerable advantage. And I believe the main objective of British energy policy is seeing that we exploit and take the fullest advantages of these unique opportunities.

[5]

Discussion

PROFESSOR N. KURTI, F.R.S. (*Clarendon Laboratory, Parks Road, Oxford*)

I was very glad to hear the Secretary of State emphasize the importance of good thermal insulation and say that standards in new buildings will be improved. However, I do not share his pessimism about improving the thermal insulation in *existing* dwellings. Techniques for double glazing, improved insulation for cavity walls, internal claddings for walls already exist, and given reasonable incentives many householders and owners of business and industrial premises would employ these techniques and thereby save the country much fuel.

Phil. Trans. R. Soc. Lond. A. **276**, 413–430 (1974)
Printed in Great Britain

WORLD ENERGY RESOURCES AND DEMAND

The world energy outlook to the mid-1980s: the effect of an alternative supply path in the United States

By J. DARMSTADTER AND S. H. SCHURR*

Resources for the Future, Inc., Washington, D.C., U.S.A.

A brief review and interpretation of regional and world-wide trends in total energy consumption and its composition since the end of World War II is given. A review of energy-consumption projections into the 1980s – world-wide and regional – focuses on the role of international trade in oil in achieving supply–demand balances. The prospective position of the U.S. as a major oil importer is emphasized. An analysis of the sensitivity of world supply prospects to alternative assumptions concerning the growth of indigenous sources of supply in the United States of America and Western Europe is presented.

The post-war growth rate in world energy consumption averaged out to over 5% per annum. Marked shifts in regional shares and variations in regional growth rates have occurred, but regional differences in the level of per capita energy use, while narrowing, remain conspicuously wide. The sharp relative decline of coal during this period was accompanied by a dramatic relative increase in both oil and gas. The rapid growth of world energy consumption as a whole, the continued shift toward oil and the rising volume of U.S. oil imports all failed to be adequately anticipated in past energy projections.

A standard projection to the mid-1980s shows: world-wide energy growth of between $5\frac{1}{2}$ and 6%; an even faster growth rate for oil, resulting in about 115×10^6 barrels (18.3×10^6 m³)/day in 1985 (compared to 53×10^6 b (8.4×10^6 m³)/d in 1972); and the addition of the U.S. to the ranks of the major oil importers.

The Middle East, along with areas of lesser reserve holdings, is in all likelihood physically capable of accommodating expected oil demand to the mid-1980s. But the acute degree of dependence that this would pose for major consuming regions prompts the question of how a greatly expanded indigenous producing capability in the U.S. could blunt the one-sidedness of the demand–supply picture. Recently completed research suggests that, within an appropriate policy setting, the U.S. could probably meet all but 20% of its oil and gas internally by 1985 – and do so at real prices no higher than the $6/barrel ($38/m³) delivered price rapidly being approached by Persian Gulf crude. Such a development, along with whatever contribution can be made by Western Europe's own petroleum-producing capability, can perhaps introduce a stabilizing element of major importance into world energy flows.

This paper begins by describing regional and world-wide trends in energy consumption since the end of World War II. This description is followed by a brief review of some past energy projections and an assessment of their accuracy in estimating future developments. In the past, there has been repeated underestimation of growth in total energy and, especially, oil demand. The conventional demand–supply picture to the mid-1980s is presented next, particularly as it bears on the emergence of the United States as a potentially large-scale importer of foreign oil, alongside such already important buyers as Western Europe and Japan. Finally, we want to re-examine world demand and supply prospects within this conventional picture of the future in the light of an alternative assumption concerning the growth of indigenous sources of supply in the United States, and in a much more cursory treatment, Western Europe. The importance of such an evaluation is underscored by the events of October and November 1973, which have once again raised questions about the risks of sustained reliance upon Middle East–North African suppliers for large volumes of petroleum.

* Present address: The Electric Power Research Institute, Palo Alto, California, U.S.A.

PAST TRENDS

World-wide energy consumption reached as estimated level of 250×10^{18} J (237×10^{15} Btu) in 1972 – equivalent to 18.4×10^6 m³/day oil. (Some 46 % of this total, or 8.4×10^{12} m³/day, was actually accounted for by oil.) The postwar growth rate in world energy consumption averaged out to over 5 % annually, and during the past decade reached about $5\frac{1}{2}$ %. At 2 % per annum population growth, this has meant a yearly growth in per capita energy consumption of approximately $3\frac{1}{2}$ % (see table 1).

Some rather striking post-World War II geographic shifts have accompanied the growth in total world energy utilization. In a number of areas, these shifts represented a continuation of trends that were in progress even earlier. For example, the Soviet Union's share of world energy consumption stood at under 2 % in 1925; just before World War II it was up to 10 %,

TABLE 1. WORLD ENERGY CONSUMPTION AND POPULATION, SELECTED YEARS, 1925–72

year	total energy consumption		population (million)	energy consumption per capita	
	10^{18} J	10^{12} Btu		10^9 J	10^6 Btu
1925	47	44249	1890.1	24.6	23.4
1950	81	76823	2504.5	32.4	30.7
1955	105	99658	2725.6	38.6	36.6
1960	131	124046	2989.9	43.8	41.5
1965	169	160722	3281.2	51.6	49.0
1968	200	189737	3484.5	57.5	54.5
1970	226	214496	3608.6	62.6	59.4
1971	235	223522	3678.3	64.2	60.8
1972	250	237166	3747.2	66.8	63.3

average annual percentage rates of change

1925–50	2.2		1.1		1.1
1950–55	5.3		1.7		3.6
1955–60	4.5		1.9		2.5
1960–65	5.3		1.9		3.4
1965–70	5.9		1.9		3.9
1970–72	5.2		1.9		3.2
1950–60	4.9		1.8		3.1
1960–70	5.6		1.9		3.6
1960–72	5.5		1.9		3.5
1950–70	5.3		1.8		3.4
1950–72	5.3		1.8		3.4

Notes and Sources for tables 1–6. Data for 1968 and years preceding are taken from Sam H. Schurr (ed.), *Energy, economic growth and the environment* (Baltimore: Johns Hopkins University Press for Resources for the Future, 1972), pp. 177–87, or, in a few cases, from sources described there.

Post-1968 energy consumption was estimated on the basis of data taken from U.S. Bureau of Mines releases (for the United States); and, for other countries, from United Nations sources [*Statistical yearbook*, 1972 (New York, 1973), *World energy supplies* 1961–1970, series J, no. 15 (New York, 1972)], and British Petroleum Co., *Statistical review of the world oil industry* 1972 (1973). Population figures come from the cited two U.N. sources and (estimated for 1972 on the basis of incomplete data) from U.N. *Monthly bulletin of statistics* (July 1973).

The post-1968 energy data in the foregoing sources were consulted as to *changes* (by energy source and region) rather than adopted as to *absolute level*. The 1968 figures, to which these changes were linked, controlled as to level.

Two characteristics of the energy consumption measure used in these tables should be noted: (1) bunker fuel and non-energy uses are included; (2) the value of primary electricity (hydro, nuclear, geothermal) was calculated, not by the heat value produced, but rather by the estimated (higher) fuel inputs required at fossil-fueled thermal electric power stations.

and from its early postwar share of around 11% it rose to approximately 15% in 1970 (see table 2).†

Rising shares occurred elsewhere in the world in the past several decades: the other Communist countries, Latin America, Africa, and Asia all exhibit long-term increases in their relative standing. The postwar rise of Asia reflects in part the growing share of the region's developing countries; but to an even greater extent it reflects the phenomenal momentum of energy growth in Japan, whose annual postwar rates of over 10% have been sustained to the most recent years for which figures are available.

TABLE 2. WORLD ENERGY CONSUMPTION AND POPULATION (%), BY MAJOR REGION, PERCENTAGE DISTRIBUTION, 1950, 1960, 1970, AND AVERAGE ANNUAL PERCENTAGE RATES OF CHANGE, 1960–70

region	percentage distribution						average annual percentage rates of change, 1960–70		
	1950		1960		1970				
	energy consumption	population	energy consumption	population	energy consumption	population	energy consumption	population	energy consumption per capita
North America	48.0	6.6	39.3	6.6	34.7	6.3	4.3	1.3	3.0
Canada	3.5	0.5	3.1	0.6	3.3	0.6	6.1	1.8	4.2
United States	44.5	6.1	36.1	6.0	31.4	5.7	4.2	1.3	2.9
Western Europe‡	22.8	12.1	21.0	10.9	22.3	9.9	6.3	0.9	5.3
Oceania	1.2	0.5	1.1	0.5	1.1	0.5	5.8	2.2	3.5
Latin America	3.1	6.5	4.0	7.1	4.3	7.8	6.3	2.9	3.4
Asia (excl. Communist)	5.0	32.2	6.6	32.5	9.7	34.1	9.7	2.4	7.1
Japan	2.3	3.3	3.0	3.1	5.2	2.9	11.9	1.0	10.7
Other Asia‡	2.7	28.9	3.7	29.4	4.5	31.3	7.7	2.5	5.0
Africa	1.7	8.7	1.7	9.2	1.7	9.7	5.4	2.4	3.0
U.S.S.R. and Comm. Eastern Europe	16.7	10.8	20.9	10.5	20.9	9.7	5.6	1.1	4.5
U.S.S.R.	11.0	7.2	14.4	7.2	14.9	6.7	6.0	1.2	4.7
Eastern Europe‡	5.7	3.6	6.5	3.3	6.0	2.9	4.7	0.7	4.0
Communist Asia	1.6	22.8	5.3	22.7	5.3	22.0	5.6	1.6	3.9
World	100.0	100.0	100.0	100.0	100.0	100.0	5.6	1.9	3.6

Source. See table 1.

‡ Yugoslavia is included in Western Europe throughout these tables; Turkey appears in Other Asia.

A principal offset to these long-term increases in regional shares of world-wide energy consumption has been the declining relative position of North America. Although North America retains its leading world share, with the United States occupying the top-ranking country share, its proportion of world energy consumption (which had been as high as 50% in the mid-1920s) fell from 45% in 1950 to under 33% in 1970. The virtually unchanged West European share of world energy consumption – at 22% since 1950 – reflects the fact that in 1950 the area was still in the midst of postwar recovery; in 1925 its world share had been around 35%.

These distributional shifts in total energy consumption among different parts of the world have also been accompanied by moderately disparate trends in the growth of per capita

† For a review of pre-World War II trends, see J. Darmstadter *et al.*, *Energy in the world economy* (Baltimore: Johns Hopkins University Press for Resources for the Future, 1971), especially Part One.

TABLE 3. WORLD ENERGY CONSUMPTION AND POPULATION, BY MAJOR REGIONS, 1950, 1960, 1970

region	1950 total consumption (10^{18} J)	1950 total consumption (10^{12} Btu)	1950 population (million)	1950 consumption per capita (10^9 J)	1950 consumption per capita (10^6 Btu)	1960 total consumption (10^{18} J)	1960 total consumption (10^{12} Btu)	1960 population (million)	1960 consumption per capita (10^9 J)	1960 consumption per capita (10^6 Btu)	1970 total consumption (10^{18} J)	1970 total consumption (10^{12} Btu)	1970 population (million)	1970 consumption per capita (10^9 J)	1970 consumption per capita (10^6 Btu)
North America	38.8	36 860	166.1	235	221.9	51.3	48 701	198.7	258	245.1	78.5	74 483	226.2	347	329.3
Canada	2.85	2 707	13.7	208	197.6	4.1	3 885	17.9	229	217.0	7.4	7 039	21.4	347	328.9
United States	36.0	34 153	152.3	236	224.3	47.4	44 816	180.7	262	248.0	71	67 444	204.8	347	329.3
Western Europe	18.4	17 483	302.4	61	57.8	27.4	26 066	326.5	84	79.8	50.5	47 874	356.4	143	134.3
Oceania	0.94	890	12.2	77	73.0	1.48	1 398	15.4	96	90.8	2.6	2 451	19.2	135	127.7
Latin America	2.52	2 397	161.9	15.6	14.8	5.2	4 939	212.4	25	23.3	9.6	9 130	282.0	34	32.4
Asia (excl. Comm.)	4.01	3 804	805.4	5	4.7	8.7	8 228	970.6	9	8.5	22	20 819	1231.3	18	16.9
Japan	1.83	1 739	82.9	22	21.0	3.9	3 672	93.2	41.5	39.4	12	11 255	103.4	116	108.9
Other Asia	2.17	2 063	722.5	3	2.9	4.8	4 556	877.4	5.5	5.2	10	9 564	1127.9	9	8.5
Africa	1.37	1 297	217.0	6.3	6.0	2.3	2 162	276.0	8.2	7.8	3.9	3 677	349.5	11	10.5
U.S.S.R. and Comm. Eastern Europe	13.5	12 842	269.8	50.2	47.6	27.4	25 973	312.9	87.5	83.0	47.2	44 758	348.4	136	128.5
U.S.S.R.	8.9	8 427	180.0	49.4	46.8	18.9	17 898	214.4	88	83.5	33.8	31 994	242.8	139	131.8
Eastern Europe	4.65	4 414	89.7	52	49.2	8.5	8 075	98.5	86.5	82.0	13.5	12 764	105.6	128	120.9
Communist Asia	1.32	1 250	569.8	2.3	2.2	6.9	6 579	677.5	10.2	9.7	12	11 304	795.6	15	14.2
World	81	76 823	2504.5	32.4	30.7	131	124 046	2989.9	43.8	41.5	226	214 496	3608.6	63	59.4

Source. See table 1.

[10]

energy consumption, both during the long-range time-span since the 1920s and during the briefer recent period highlighted in table 2. During the 1960–70 decade, for example, per capita energy growth ranged from around 3 % in the U.S. to over 10 % in Japan, with a number of regions clustered in the 4–5 % per capita growth rate bracket.

But regional differences in the level of per capita energy use, while narrowing, remain dramatically wide, as can be seen in table 3. In 1970 U.S. per capita energy consumption of 347×10^9 J and that of Canada, which was only slightly lower, were more than two and one-half times the level of the next ranking regions – Western Europe, the U.S.S.R., and Oceania, all of which recorded per capita figures in the vicinity of 137×10^9 J. And the more extreme disparity is reflected in the fact that North American per capita consumption was between 30 and 40 times the levels prevailing in Africa and the developing portions of Asia.

If North America's level of per capita energy consumption in 1925 (roughly 185×10^9 J) had remained unchanged throughout the period 1925–70, it would still have been some 30 % above the next highest area (Western Europe) tabulated for the year 1970 and nearly 20 times as high as the developing countries of Asia and Africa. To be sure, trends and levels in per capita energy consumption are not synonymous with per capita income or gross national product; nor are the latter measures, in turn, truly reflective of living standards, however defined. None the less, there is unquestionably a sufficiently close connexion between levels of per capita energy consumption and general economic development to permit one to point to the more extreme disparaties of table 3 as yet another sign that substantial improvement in living standards in the years ahead will constitute a growing burden on world energy supplies.

TABLE 4. WORLD ENERGY CONSUMPTION: PERCENTAGE DISTRIBUTION BY SOURCE 1950–72

	1950	1960	1965	1968	1970	1971	1972
coal†	55.7	44.2	39.0	33.8	31.2	29.9	28.7
oil‡	28.9	35.8	39.4	42.9	44.5	45.2	46.0
natural gas	8.9	13.5	15.5	16.8	17.8	18.3	18.4
primary electricity§	6.5	6.4	6.2	6.5	6.5	6.6	6.9
total	100.0	100.0	100.0	100.0	100.0	100.0	100.0
10^{15} J	80	130	168	200	226	236	250
10^{18} Btu	76.8	124.0	160.7	189.7	214.5	223.5	237.2

Source. See table 1.

† Principally bituminous coal, but also including anthracite, lignite, and a variety of low-quality coals.

‡ Including, where known, natural-gas liquids.

§ For 1971, the 'primary electricity' components were as follows [Joules converted at the fuel-input equivalent (see footnote to table 1)]:

	10^9 kW h	10^{15} J	as % of world energy consumption
geothermal	5.6	64	0.03
nuclear	103.9	1195	0.5
hydro	1243.3	14300	6.1

Tables 4, 5 and 6 provide a picture of postwar shifts in the role of the different energy sources. The sharp relative decline of coal during this period was accompanied by marked relative increases in both oil and gas. The latter two accounted for 38 % of world-wide energy consumption in 1950 and 64 % in 1970; concurrently, coal experienced a relative decline, from 56 to 29 %. The primary electricity share – hydro along with a thus far modest but rising nuclear component – remained essentially unchanged.

[11]

To a greater or lesser extent, similar changes – at least the shift from coal to oil and gas – occurred in principal regions of the world. In no area did the share of coal fail to decline. In each area, the share of natural gas was higher in 1970 than in 1950. This was true also of oil consumption in each area, with the exception of Latin America; in that region (where oil had already constituted 73% of energy consumption in 1950) the oil proportion dropped somewhat, while that for gas rose sharply. Only in two regions – Eastern Europe and China – did coal continue to contribute more than half of total energy consumption in 1970. In Western Europe, not only was there a sharp decline in coal's relative share in less than two decades from over

TABLE 5. WORLD ENERGY CONSUMPTION: AVERAGE ANNUAL PERCENTAGE RATES OF CHANGE, BY SOURCE, SELECTED PERIODS, 1950–72

	1950–60	1960–65	1965–70	1970–72	1960–70	1950–70	1960–72	1950–72
coal	2.5	2.8	1.4	0.9	2.0	2.3	1.8	2.1
oil	7.1	7.4	8.6	6.9	8.0	7.6	7.8	7.5
natural gas	9.4	8.3	8.9	6.8	8.6	9.0	8.3	8.8
primary electricity	4.8	3.9	7.6	8.3	5.7	5.2	6.2	5.5
total	4.9	5.3	5.9	5.2	5.6	5.3	5.6	5.3

Source. See table 1 and footnotes to table 4.

three-fourths to little over one-fourth, but its absolute use also fell. Concurrently, oil and gas went from 15 to 62%. Japan's energy pattern disclosed similarly dramatic shifts. The Soviet picture is highlighted by a big postwar rise in the share of natural gas – from 2 to 22%. The changing U.S. pattern was far less remarkable, for the early pre-eminence of the United States in world oil and gas resulted in important shares for these fuels in the country's total energy consumption far earlier than in most other regions. Thus, the proportion of oil in U.S. energy consumption rose rather modestly between 1950 and 1970, the sharply declining relative position of coal (with, incidentally, only slight long-term absolute growth) being principally compensated for by natural gas. In the past several years, however, U.S. oil consumption has advanced at a disproportionately fast rate as gas output levelled off, power-station coal utilization encountered tight environmental constraints, and oil demand in the transport sector accelerated.

This dependence on oil, as a 'balancing' element in U.S. energy requirements, has given rise to substantially expanded U.S. oil imports during the past ten years, as table 7 shows. Rising from 334×10^3 m³/day in 1962 to 747×10^3 m³/day in 1972 – or at a growth rate of $8\frac{1}{2}$% per year – the oil import share of U.S. oil consumption has gone from 20 to 30% (and earlier this year was running at close to 35%). In terms of national energy consumption in the aggregate, however, U.S. oil imports stand far below comparable shares elsewhere. In Western Europe, oil imports went from 37% of total energy consumption in 1962 to nearly 60% in 1972; in Japan, from 44 to 73%; while, for the United States, the increase was from 9 to $13\frac{1}{2}$%.

REVIEW OF PROJECTIONS

Three principal lines of development discussed earlier – the rapid growth of world energy consumption as a whole, the continued shift towards oil, and the rising volume of U.S. oil imports – all failed to be adequately anticipated in the succession of energy projections which have appeared since around 1960. The forecasting record, summarized in table 8, is designed for impressionistic purposes only. A retrospective analysis comparing the different projection

TABLE 6. PERCENTAGE DISTRIBUTION OF ENERGY SOURCES FOR MAJOR REGIONS, 1950 AND 1970

10^{18} J	1950 coal	oil	natural gas	primary electricity	%	total 10^{18} J	total 10^{15} Btu	1970 coal	oil	natural gas	primary electricity	%	total 10^{18} J	total 10^{15} Btu
North America	38.0	38.7	16.7	6.4	100.0	38.8	36.86	18.3	43.7	32.1	6.0	100.0	78.5	74.48
Canada	40.6	28.6	2.8	27.9	100.0	2.85	2.71	9.6	41.3	26.9	22.2	100.0	7.4	7.04
United States	37.8	39.5	18.0	4.7	100.0	36.0	34.15	19.1	43.9	32.7	4.3	100.0	71	67.44
Western Europe	77.4	14.3	0.3	8.0	100.0	18.4	17.48	27.4	55.6	6.1	10.8	100.0	50.5	47.87
Oceania	65.3	27.3	—	7.4	100.0	0.94	0.89	39.8	48.6	2.2	9.3	100.0	2.6	2.45
Latin America	9.8	72.9	8.3	9.0	100.0	2.52	2.40	4.9	67.8	18.4	8.8	100.0	9.6	9.13
Asia (excl. Comm.)	53.3	28.5	1.4	16.9	100.0	4.01	3.80	23.3	64.1	6.2	6.4	100.0	22	20.82
Japan	61.9	5.0	0.2	32.9	100.0	1.83	1.74	22.4	68.8	1.3	7.5	100.0	12	11.26
Other Asia	46.0	48.3	2.4	3.3	100.0	2.17	2.06	24.3	58.7	11.8	5.2	100.0	10	9.56
Africa	61.4	36.9	—	1.7	100.0	1.37	1.30	43.5	48.7	1.5	6.2	100.0	3.9	3.68
U.S.S.R. and Comm.	81.4	14.6	2.3	1.7	100.0	13.5	12.84	49.6	28.7	18.5	3.1	100.0	47.2	44.76
Eastern Europe														
U.S.S.R.	75.6	19.7	2.5	2.3	100.0	8.9	8.43	40.4	33.3	22.5	3.8	100.0	33.8	31.99
Eastern Europe	92.5	4.8	2.0	0.7	100.0	4.65	4.41	72.7	17.4	8.4	1.5	100.0	13.5	12.76
Communist Asia	92.8	0.9	—	6.3	100.0	1.32	1.25	89.4	8.2	NA†	2.4	100.0	12	11.30
World	55.7	28.9	8.9	6.5	100.0	81	76.82	31.2	44.5	17.8	6.5	100.0	226	214.50
10^{18} J	45.1	23.4	7.19	5.28		100.7		70.6	100.7	40.3	14.7			
10^{15} Btu	42.80	22.20	6.82	5.00				66.90	95.48	38.21	13.91			

Sources. See table 1 and footnotes to table 4.

† The UN (basic source for these data; see notes to table 1) shows no figures on China's natural-gas production and consumption. A recent analysis puts China's natural-gas production at 12.1×10^9 m³ (417×10^{15} J; 0.4×10^{15} Btu) in 1965 and 34.0×10^9 m³ (1.25×10^{18} J; 1.2×10^{15} Btu) in 1971 (C. S. Chen and K. N. Au, The Petroleum industry of China, *Die Erde*, Heft 3–4, 1972 (Berlin), p. 319). Such a quantity would clearly raise the estimate of China's energy consumption – shown above and in tables 2 and 3 – quite markedly. Another estimate (*Oil and Gas J.* 20 August 1973) credits China with only about 20% of this amount of gas output.

TABLE 7. ENERGY CONSUMPTION, OIL CONSUMPTION, AND OIL IMPORTS: UNITED STATES, WESTERN EUROPE AND JAPAN, 1962 AND 1972

	1962			1972		
	United States	Western Europe	Japan	United States	Western Europe	Japan
	10^6 m³/day (10^6 barrels/day)					
energy consumption (oil equivalent)	3.7 (23.27)	2.22 (13.96)	0.358 (2.25)	5.57 (35.05)	3.79 (23.84)	1.05 (6.58)
oil consumption	1.63 (10.23)	0.832 (5.24)	0.153 (0.96)	2.54 (15.98)	2.25 (14.20)	0.762 (4.80)
oil imports†	0.335 (2.12)	0.825 (5.19)	0.156 (0.98)	0.75 (4.74)	2.23 (14.06)	0.76 (4.78)
from Middle East–North Africa‡	0.054 (0.34)	0.604 (3.80)	0.115 (0.72)	0.11 (0.70)	1.80 (11.30)	0.60 (3.78)
from elsewhere	0.283 (1.78)	0.22 (1.39)	0.041 (0.26)	0.64 (4.04)	0.44 (2.76)	0.16 (1.00)
	percentage of energy consumption					
oil consumption	44.0	37.5	42.7	45.6	59.6	73.0
oil imports†	9.1	37.2	43.6	13.5	59.0	72.6
from Middle East–North Africa‡	1.5	27.2	32.0	2.0	47.4	57.4
from elsewhere	7.6	10.0	11.6	11.5	11.6	15.2
	percentage of oil consumption					
oil imports†	20.7	99.0	102.1	29.7	99.0	99.6
from Middle East–North Africa‡	3.3	72.5	75.0	4.4	79.5	78.6
from elsewhere	17.4	26.5	27.1	25.3	19.4	20.9
	percentage of oil imports					
from Middle East–North Africa‡	16.0	73.2	73.5	14.9	80.4	78.9
from elsewhere	84.0	26.8	26.5	85.1	19.6	21.1

Sources. Data for 1962 based on Joel Darmstadter *et al.*, *Energy in the world economy* (Baltimore: Johns Hopkins University Press for Resources for the Future, 1971), and British Petroleum Co., *Statistical review of the world oil industry*, 1962; data for 1972 based on British Petroleum Co., *Statistical review of the world oil industry*, 1972.

† Imports are gross of exports. Thus, they exclude product exports from West European refineries. And for Japan, excess of imports over consumption arises because of small quantities of product exports, refinery losses and (presumably) independent construction of the two series. By showing gross rather than net imports, we overstate slightly the degree of foreign dependence. The overstatement matters, if at all, only in the case of West Europe.

‡ Includes negligible quantities from West Africa in 1962.

studies as well as a comparison between the projections and actual performance cannot be adequately detailed here.§

This said, the degree of underestimation disclosed by the table is still worth pondering. As we might expect, there is less error in total energy projections than in those on oil. But even for the total the projections have been markedly conservative. In the U.S. case, which we have dissected in some detail elsewhere,‖ the progressively declining relationship – observed for decades

§ Comparability is impaired for a number of reasons: definitions do not always agree; a varying amount of historical information may have been available for the various studies, even though ultimately published in the same year; projected time-intervals do not precisely correspond to the actual time lapse used in the comparison. Moreover, assumptions governing the different forward estimates – both as to total energy and particular fuels – would need to be standardized or, at least, made explicit in order to permit a diagnosis of what went wrong and for what reason.

‖ Sam H. Schurr (ed.), *Energy, economic growth, and the environment* (Baltimore: Johns Hopkins University Press for Resources for the Future, 1972), pp. 168–171.

– between energy growth and g.n.p. growth did not endure into the later part of the 1960s as had been assumed. What was not foreseen, among other things, was the halt to efficiency improvements in the U.S. electric power sector and the acceleration in demand growth for motor vehicle fuels. Since 1970 the U.S. energy/g.n.p. ratio seems once again to have reverted to its long-term downward trend. It does seem, however, from the rather erratic energy/g.n.p.

TABLE 8. REVIEW OF SELECTED PAST ENERGY CONSUMPTION PROJECTIONS

	actual data†			projections		
region	period	average annual growth rate (%)	source	year published	period	average annual growth rate (%)
			total energy consumption			
World	1960–70	5.6	(A)	1966	1960–70	4.6
					1970–80	4.8
			(B)	1966	1960–80	5.0
Western Europe	1960–70	6.3	(C)	1960	1955–75	2.8
			(A)	1966	1960–70	4.4
					1970–80	4.0
			(B)	1966	1964–70	4.2
					1970–80	4.1
United States	1960–70	4.2	(D)	1963	1960–70	2.9
					1970–80	2.8
			(A)	1966	1960–70	3.6
					1970–80	3.3
Japan	1960–70	11.9	(A)	1966	1960–70	9.1
					1970–80	7.0
			(B)	1966	1964–70	10.0
					1970–80	6.9
			oil consumption			
Western Europe	1962–72	10.5	(B)	1966	1964–80	4.1‡
United States	1962–72	4.6	(A)	1966	1960–70	3.5
					1970–80	2.7
			(E)	1971	1965–75	3.4
					1975–80	2.9
			(F)	1968	1965–80	3.1
Japan	1962–72	17.5	(B)	1966	1964–70	14.3
					1970–80	8.3
			oil imports			
United States	1962–72	8.4	(F)	1968	1965–80	3.2
			(E)	1971	1965–75	3.5
					1975–80	2.6
			(A)	1966	1960–70	4.4
					1970–80	4.2§

Sources. (A) European Coal and Steel Community, *Review of the long-term energy outlook of the European Community.* (B) O.E.C.D., *Energy policy,* (C) O.E.E.C., *Towards a new energy pattern in Europe.* (D) Hans Landsberg, Leonard L. Fischman and Joseph L. Fisher, *Resources in America's future* (Baltimore: Johns Hopkins Press for Resources for the Future). (E) Sam H. Schurr, Paul T. Homan and associates, *Middle Eastern oil and the western world: prospects and problems* (New York: American Elsevier Publishing Co.). (F) U.S. Department of the Interior, *United States petroleum through 1980.*

Note. Differences in definitional practices among the various projection studies impairs exact comparison, even of growth rates.

† Taken from tables 1 or 7 above.
‡ Highest of range shown in source (B).
§ Midpoint of range shown in source (A).

[15]

elasticities tabulated for a number of countries (see table 9), that this familiar (and perhaps over-used) relationship may have limited utility as a basis for forecasting energy consumption.

More spectacular misjudgements, evident in table 8, concern the role of oil in Western Europe's energy balance and of oil consumption and imports into the United States. In the case of recent trends in Western Europe's oil consumption, a key element seems to lie in the fact that the 1966 O.E.C.D. study, *Energy policy*, allowed for a far greater role for coal between 1964 and 1970 (though not an expanding one) than actually took place. In this respect the 1966 O.E.C.D. study was only the latest in a succession of projections which failed to anticipate the rapidly shrinking role of the West European coal industry.

TABLE 9. ENERGY CONSUMPTION GROWTH RELATIVE TO G.N.P. GROWTH, SELECTED COUNTRIES, 1960–71

	1960–65 average annual % rate of change in:			1965–71 average annual % rate of change in:		
	(1) energy consumption	(2) real g.n.p.	(3) elasticity (1)/(2)	(1) energy consumption	(2) real g.n.p.	(3) elasticity (1)/(2)
United States	3.9	4.8	0.81	4.5	3.1	1.45
Canada	6.2	5.6	1.11	6.1	4.8	1.27
United Kingdom	2.5	3.3	0.76	1.6	2.5	0.64
France	5.5	5.8	0.95	5.7	5.7	1.00
West Germany	5.8	5.0	1.16	4.2	4.6	0.91
Italy	9.1	5.3	1.72	7.8	5.2	1.50
Sweden	5.8	4.9	1.18	6.1	3.3	1.85
Switzerland	6.8	5.3	1.28	6.0	3.9	1.54
Spain	5.8	8.6	0.67	8.9	6.0	1.48
Belgium–Luxembourg	6.1	4.8	1.27	4.9	4.6	1.06
Netherlands	7.8	5.0	1.56	7.9	5.3	1.49
Denmark	8.6	5.2	1.65	5.2	4.4	1.18
Japan	10.0	10.0	1.00	11.8	11.2	1.05
Australia	6.3	5.0	1.26	4.6	5.1	0.90
India	7.0	3.3	2.12	3.5	4.2	0.83
Venezuela	5.7	7.7	0.74	2.6	4.2	0.56
Mexico	4.6	7.0	0.66	7.9	6.3	1.25

Sources. Data for 1960–5 from Darmstadter, *Energy in the world economy, op. cit.*; energy data 1965–71 from United Nations, *World Energy Supplies*, series J, various numbers; GNP data, 1965, from U.S. Agency for International Development, 'Gross National Product – growth rates and trend Data by Region and Country', Release RC-W-138, 1 May 1973.

Note. The energy consumption growth rates shown here may differ slightly from figures in other tables because of differences in derivation.

As for U.S. oil demand and imports, we have already referred to some of the critical factors at work in that development: severe environmental restrictions against the use of coal, lagging natural gas output, and accelerated demand growth in the transport sector, all put pressure on oil demand, which, given the levelling off in America's own oil production, have contributed to a totally unforeseen need for foreign oil in the quantities which have been entering the country.

THE CONVENTIONAL INTERNATIONAL PICTURE FOR THE MID-1980s

Two standard representations of future energy developments, which have recently become available, are those of the National Petroleum Council (in the United States) and the Organization for Economic Cooperation and Development.† We have depended upon these studies (modified, to some extent) as well as upon several subsidiary analyses in fashioning the summary projections of tables 10 and 11.‡

TABLE 10. ILLUSTRATIVE PROJECTIONS OF WORLD ENERGY CONSUMPTION,
BY SOURCE, 1972–85

	1972			1972–85 average annual % rate of change	1985		
	10^{15} J/day	oil equivalent (10^6 b/d)	%		10^{15} J/day	oil equivalent (10^6 b/d)	%
total	709	116	100.0	5.7	1460	239	100.0
coal	202	33	28.7	2.4	275	45	18.8
oil	324	53	46.0	6.2	709	116	48.5
gas	128	21	18.4	6.4	287	47	19.7
primary electricity	49	8	6.9	11.0	190	31	13.0
(hydro)	(43)	(7)	(6.1)	(6.0)	(92)	(15)	(6.3)
(nuclear)	(6)	(1)	(0.8)	(24.0)	(98)	(16)	(6.7)

Sources. Data for 1972 from tables 1 and 4 (and sources thereto) converted from Btu's into barrels/day oil equivalents. 1985 by assumption, but representing in large part an adaptation, with some modifications, of projections to 1980 in O.E.C.D., *Oil, the present situation and future prospects* (Paris, 1973). Data shown conform to concepts described in notes to table 1. See text for additional comments.

Four central features of the standard representation of projected trends in world energy consumption to the mid-1980s are: first, a continued rapid world-wide growth of between $5\frac{1}{2}$ and 6 % per annum in total energy demand; secondly, a correspondingly rapid – indeed somewhat faster – expansion in oil demand, resulting in world-wide consumption of about 18.4×10^6 m³/day in 1985 compared to around 8.4×10^6 m³/day in 1972 (a continuously declining coal share is offset by comparatively fast growth for nuclear energy and natural gas); thirdly, perhaps the most fundamental change over the past patterns of world energy flows which is presently foreseen is the sharply rising volume of U.S. oil imports – up from 800×10^3 m³/day in 1972 to about 2.1×10^6 m³/day in 1985; and, fourthly, in spite of what, in historical terms, is a significant expansion of Western Europe's indigenous oil-producing capability, the pre-eminent share of that region's oil needs (and, of course, that of Japan also) will have to be met by imports.

It is assumed in this projection that the growth in energy requirements in relationship to that in economic growth will generally remain at or above unity for the major consuming regions, and substantially above unity elsewhere in the world. This implies that the real price of oil, while it can reasonably – some would say inescapably – be expected to rise during the next

† National Petroleum Council, *U.S. Energy outlook* (Washington, December 1972); and Organization for Economic Cooperation and Development, *Oil – the present situation and future prospects* (Paris, 1973).

‡ A literal adoption, jointly of the N.P.C. and O.E.C.D. projections, is impossible since there are some differences in definitions, assumptions, and time-periods surveyed; there is also the fact that the N.P.C. postulates a number of alternative cases while the O.E.C.D. works with a single set of projected figures.

TABLE 11. ILLUSTRATIVE PROJECTIONS OF WORLD OIL CONSUMPTION AND POSSIBLE
PRODUCTION LEVELS, SELECTED AREAS, 1972–1985

	1972†		1985	
	$10\ m^3/d$	$10^6\ b/d$	$10\ m^3/d$	$10^6\ b/d$
consumption				
United States	2.54	15.98	4.10	25.8
Canada	0.26	1.66	0.48	3.0
Western Europe	2.26	14.20	4.50	28.3
Japan‡	0.76	4.80	2.32	14.6
U.S.S.R., China, Eastern Europe	1.27	7.99	3.29	20.7
Other	1.28	8.07	3.75	23.6
total	8.37	52.70	18.4	116.0
actual (1972) and possible production levels				
United States	1.78	11.18	1.96	12.3
Canada	0.29	1.84	0.71	4.5
Western Europe	0.07	0.44	0.64	4.0
Latin America	0.79	4.98	1.18	7.4
Middle East	2.86	17.98	7.79	49.0
North Africa	0.59	3.74	1.07	6.7
West Africa	0.33	2.08	0.98	6.2
U.S.S.R., China, Eastern Europe	1.41	8.87	3.29	20.7
Indonesia, Australia, other Eastern Hemisphere	0.29	1.83	0.83	5.2
total	8.41	52.92	18.4	116.0
net import requirements				
United States	0.76	4.80	2.14	13.5
Western Europe	2.18	13.76	3.86	24.3
Japan‡	0.76	4.80	2.32	14.6
total	3.70	23.36	8.32	52.4
balancing production and net exports				
production:				
Middle East	2.86	17.98	7.79	49.0
North Africa	0.59	3.74	1.07	6.7
West Africa	0.33	2.08	0.98	6.2
Latin America	0.79	4.98	1.18	7.4
Indonesia, Australia, other Eastern Hemisphere	0.29	1.83	0.83	5.2
total	4.86	30.61	11.85	74.5
less: own consumption, above areas	1.28	8.07	3.75	23.6
equals: net exports, above areas	3.58	22.54	8.10	50.9
plus: net exports, Canada	0.0286	0.18	0.24	1.5
net exports, U.S.S.R., China, Eastern Europe	0.14	0.88	—	—
equals: net exports	3.75	23.60	8.34	52.4

Source. Data for 1972 from British Petroleum Co., *Statistical review of world oil industry*, 1972. Projections for 1985: *total* consumption (and production) taken from table 10. Distribution by consuming and producing countries, using a variety of sources (in some cases modified). These included National Petroleum Council. *U.S. energy outlook* (Washington: December 1972); 'Energy: the changed and changing scene', paper by Geoffrey Chandler, the Institute of Petroleum meeting, 7 June 1973; and O.E.C.D., *Oil, the present situation and future prospects* (Paris, 1973). The U.S. figure is the N.P.C.'s Case III – involving assumption of moderately high degree of import dependence. See text for additional remarks.

† For definitional and estimating reasons, there are minor statistical discrepancies between production and consumption (and between net imports and net exports) for 1972.

‡ Japanese oil production (insignificant in 1972) is not shown separately for 1972 or considered for 1985.

decade, will none the less fail to restrain a vast expansion in oil consumption, essentially in line with past trends. World-wide (as opposed to localized) oil supply constraints are ruled out. Clearly, such assumptions – especially the last two – are highly debatable.

The net import requirements for oil of the United States, Western Europe and Japan combined are seen, in these assumptions to rise from 3.7×10^6 m³/day in 1972 to around 8.3×10^6 m³/day in 1985, implying an average annual rise of 6.5 % and a cumulative thirteen year total of about 29×10^9 m³. This total may be compared to currently published proved reserves figures for the potential net exporting regions of the world (excluding, that is, the United States, Western Europe, and the Communist countries) as follows:

	10^9 m³	10^9 barrels
Middle East: Saudi Arabia	29	183
other	27.4	172
North Africa	13.2	83
all other	12.9	81
total	82.5	519

The indispensability of future Middle East supplies, and, apparently, of Saudi Arabia within the Middle East, in accommodating this particular world demand–supply hypothesis stands out unmistakably. Substantial new reserves will of course have to be added to maintain a stable production-to-proved reserves ratio in the mid-1980s. Purposeful withholding of supplies apart, the reserves here credited to the net exporting regions are, moreover, not going to be at the exclusive disposition of the three principal importing regions. Some 12.7×10^9 m³ during 1972–85 may be consumed within the net exporting area itself.

A CONVENTIONAL VIEW OF THE U.S. POSITION IN 1985

Since it is the changed U.S. position which figures so prominently in conjecture about evolving relationships in the world oil trade, some words about the United States in particular seem called for. As already noted, most recent attempts to scan the future energy demand–supply situation in the United States have concluded that the relative U.S. dependence on imported oil seems likely to continue rising for years to come. The recently completed comprehensive report by the N.P.C. cited earlier analysed a variety of U.S. demand–supply 'scenarios' up to 1985. Total energy consumption is projected to increase at an annual growth rate of somewhat over 4 %. The N.P.C. examined a variety of alternative supply possibilities, ranging from case I, with its high domestic output potential (2.46×10^6 m³ of oil per day), to case IV – labelled 'continuation of current trends' – with its low output potential (1.65×10^6 m³ of oil per day). Most often quoted are two intermediate supply evaluations ('cases II and II'), yielding figures on oil import dependency in 1985 ranging from 38 to 53 %, or 1.43×10^6 to 2.22×10^6 m³/day. As does the N.P.C. when dealing with a single, middle of-the-range set of estimates, we have adopted case III for the illustrative purposes of table 11.

It is important to underscore the fact that the N.P.C. projections flow directly from certain implicit or explicit assumptions: growth in energy demand and compositional shifts in energy sources and forms not unlike those of recent years; minimal contribution from synthetics; and real price increases which are assumed neither to dampen demand growth nor to stimulate domestic output to levels sufficient to keep imports below the 38–53 % range indicated in the N.P.C.'s intermediate supply cases.

The prospects confronting the U.S. and the other prospective principal oil importing regions to the mid-1980s gives rise to a key question of enormous interest and concern: Does the United States (and, to a lesser extent, Western Europe) have the flexibility substantially to alter the future supply–demand outlook through purposeful measures to expand domestic output? And what would be the implications if such expandability turned out to be feasible?

ANALYSIS OF AVAILABLE U.S. OPTIONS

As part of the broader Energy Policy Project of the Ford Foundation, Resources for the Future (R.f.F.), has recently undertaken an intensive analysis of alternative energy supply possibilities for the United States. Our tentative findings confirm the view that present shortages of domestically produced energy are not the result of any shortage of undeveloped resources in the ground. The nation simply has not followed policies conducive to the development of domestic supply in the last 15–20 years. Certain policies – for example, import controls and special tax provisions – were supposed to stimulate domestic supply and no doubt they did have some effect. However, it takes a consistent package of policies to bring about increased supply and in the last 15 years the incentive of a favourable rate of return has been missing from the package. Oil and natural gas prices have been directly and indirectly kept too low. As a result, the domestic oil producing industry has actually been liquidating its assets. The drilling of oil and gas wells has lagged. In 1956, the postwar peak year, the industry drilled over 57 000 wells. Last year only 29 000 wells were drilled – the U.S. is now almost back to the 1946 level of well-drilling. This can hardly be said to be a vigorous programme of developing domestic capacity.

There are indications that the situation is improving. Crude-oil prices which in constant dollars have declined almost continuously since 1957 are turning around. Natural gas prices are beginning to increase, although natural gas, a premium fuel, in 1972 was priced, on the average, at 19 cents per 10^9 J at the wellhead, while oil was selling at 57 cents. However, natural gas prices in particular and oil prices to some extent are still subject to the hazards of regulatory, judicial, and political processes.

Various estimates exist of the amount of oil and gas remaining to be found, developed, and produced in the U.S. Industry estimates such as those made by the National Petroleum Council and in the reports of the Potential Gas Committee may well underestimate the magnitude of the resources available to the nation. Be that as it may, there does not seem to be any serious question about the adequacy of natural resources of oil and gas in the U.S. to support any reasonable degree of self-sufficiency in energy during the next few decades. Finding, developing and producing those resources will, of course, be a monumental task. As yet there is no assurance that the nation will decide to undertake the task.

Coal resources are undeniably large, although increased utilization of coal is presently hampered by serious environmental problems in both production and utilization. A policy designed to achieve substantial independence in energy supplies by the mid-1980s would require a determined research effort to develop stack gas cleaning technology for high-sulphur coal. There are indications that a reliable technology could be available, though at a considerable cost.

Resources of low-cost uranium may well be much larger than we have been led to believe by the U.S. Atomic Energy Commission figures. In any event uranium costs are such a small portion of the cost of nuclear power that the need to go to higher-cost uranium resources would not be economically disastrous to nuclear energy prospects.

It does not appear that such new technologies as coal gasification, coal liquefaction, the breeder reactor, solar energy, geothermal energy, or fusion energy will make a major contribution to commercial energy supplies before the 1990s at the earliest. For the great bulk of domestic production the U.S. will have to continue to depend on oil, natural gas, coal used as coal, and conventional reactor types – assuming their problems can be solved.

The new source and technology which is perhaps most likely to make a contribution to domestic supply within the next several decades is shale oil. Vast resources of oil shale are available in the United States to satisfy liquid-fuels demands for a very long period of time. (High-quality resources are judged of the same order of magnitude as total proved oil reserves in the Middle East.) Production costs of oil from shale cannot be known with certainty because nothing even close to a commercial-scale plant has yet been built. However, based upon extrapolation of data from small experimental plants, it appears that oil from shale could be produced in commercial plants at a cost of about \$28 to \$31/m³. This would result in delivered prices at East Coast consuming centres of around \$35 to \$38/m³.

This is a highly significant figure because, by reason of the very large domestic resource base of oil shale it sets an upper limit on the long-run cost of liquid fuels in the United States. Even if resource grade were to decline somewhat, the resulting cost-increasing pressures would likely be offset by the cost-reducing effects of improvements in the technology of extracting and processing the materials. The rate at which major crude-oil-exporting countries are increasing their prices suggests that the cost to the United States of imported crude oil clearly is heading towards levels that would make shale oil competitive.

Shale-oil production, it should however be noted, causes severe environmental problems. This may result in substantial delays in starting a commercial shale industry while ways are sought to overcome its environmental drawbacks – mainly, how to dispose of spent rock. It is worth noting that this would be the third occasion that serious consideration has been given to developing shale oil into a commercial source in the U.S. There was talk of a shale-oil industry in the 1920s and again after World War II when a modest research and development effort was launched by the government. Each time, new supplies of crude oil dimmed shale oil prospects.

Indeed, this could happen again, for it appears from the R.f.F. analysis that crude oil and natural gas may actually constitute lower-cost means for enlarging domestic supply than would oil shale. While it is unlikely that shale could make a quantitatively significant contribution to U.S. energy supplies during the next decade, there *are* encouraging prospects for the expansion of domestic crude oil and natural gas supply. Domestic natural gas offers probably the lowest-cost option for increasing domestic energy supplies and, therefore, reducing U.S. dependence on imported oil between now and the mid-1980s.

Supply difficulties in natural gas have been a basic cause of the overall energy supply problems now being experienced in the United States. U.S. government policies, now undergoing modifications, have severely limited the rise of gas prices; a severe ensuing shortage of natural gas has been aggravated by the stimulus to gas demand resulting from its attractive characteristics from an environmental standpoint. As gas supplies languished, other fuels,

[21]

particularly oil, were diverted to markets that gas would otherwise have served, leading to supply shortfalls all along the line (abetted, as noted above, by supply-restricting influences that were also affecting other fuels).

Possibilities for a substantial expansion of crude oil output at costs below those of shale oil are also indicated in the R.f.F. supply analysis. If our tentative findings are correct, it would thus appear that growth of both crude oil and natural gas production could lead to an expanded domestic supply capability in the United States, at costs ranging, say, between $31 and $38/m³ of oil (or its equivalent), delivered East Coast, and expressed in today's prices.

Moreover, these prices may well be below imported costs for Persian Gulf crude – a remarkable turnabout from conditions that have prevailed during recent years, brought on by the pricing policies now being enforced by the exporting countries. Even if these expectations with respect to conventional hydrocarbons do not materialize, ample supplies of shale oil eventually could be made available, though not without a substantial transitional degree of import dependence.

Our summary view for the United States, then, is that although the period of very low-cost energy is over, we are optimistic that long-term domestic energy prices need not increase nearly as sharply as sometimes predicted, give a favourable economic and environmental policy setting. Measured in constant 1972 dollars the R.f.F. analysis indicates that the nation can produce 80 % of its oil and natural gas requirements in 1985 at crude-oil prices not higher than $38/m³ and at natural gas prices in the $21–25/10³ m³ range. Although a $38 crude-oil price is twice the level of a few years back, it is still very low in light of the rate at which prices now seem to be increasing.

It should be noted that the stipulated degree of self-sufficiency rests on the expectation that a large expansion of U.S. natural-gas output will moderate total U.S. oil requirements to a level considerably below the consumption that is projected in the standard view high-import situation. Specifically, a reduction to 20 % import dependence in oil is assumed achievable at a 3.18×10^6 m³/day level of oil demand rather than the 4×10^6 m³/day level indicated in table 11. Also, the U.S. self-sufficiency alternative presupposes a capability to produce close to 950×10^6 tonnes of coal in 1985, compared to 600×10^6 t in 1972.

The price and quantity estimates which we have cited for future oil and gas are of course uncertain and should not be taken as predictions of the future. While we believe that the nation's resource position is good and that the basic economics of expanded domestic output are acceptable, much depends on the nation adopting the policies necessary to lead to these results on the schedule we have assumed.

A WORD ABOUT WESTERN EUROPE'S OPTIONS

Despite the strong historical position of coal in Western Europe, it appears that the most favourable prospects for significantly reducing the European level of import dependence exist, not through use of coal, but through domestic crude oil and natural gas, of which major deposits have been found only in the recent past.

An exceedingly optimistic appraisal of Western European indigenous energy has recently been developed by Professor Peter Odell.† We are in no position to subject his views to critical scrutiny but record them here because (1) they reflect the judgement of a well-qualified analyst,

† Peter Odell 1973 'Indigenous Oil and Gas in Western Europe', *Energy Policy*, **1**, 49.

and (2) they probably indicate the maximum degree of energy self-sufficiency achievable by the mid-1980s. Coupled with the R.f.F. assessment of options available to the United States, Odell's figures can thus be used to weigh the prospects for a more balanced economic relationship between the major oil exporting and importing nations. Professor Odell perceives production possibilities of indigenous oil and gas production equal to 45% of total West European energy consumption in 1985, compared to a standard projection of about an 18% share for the two fuels combined. For oil alone, the high alternative of a 50% indigenous share compares with the standard projection of around 15%.

Odell's analysis, based on a comprehensive review of the available information, leads him to the conclusion that the prospects for indigenous supplies in Western Europe have been systematically and substantially underestimated by others. Most European observers, on the other hand, feel that Professor Odell's estimates are two high. Although we are, as indicated, in no position to evaluate the validity of the figures, strictly from a physical point of view such levels of output are achievable.

Concluding Remarks

It would be piling conjecture on top of conjecture to judge how the supply alternatives which we have sketched might erode the economic leverage otherwise accruing to the major Eastern Hemisphere exporters.

If the domestic supply possibilities for the United States and Western Europe referred to above were realized, international oil flows in 1985 might look quite different:

Estimated net import requirements in 1985

	standard view		alternative view	
	10^6 m³/day	10^6 barrels/day	10^6 m³/day	10^6 barrels/day
United States	2.15	13.5	0.83	5.2†
Western Europe	3.86	24.3	1.65	10.4‡
Japan	2.32	14.6	2.32	14.6
total	8.33	52.4	4.80	30.2

† At 20% of estimated consumption, the import level of the late 1960s.
‡ Calculated from Odell (1973).

It seems safe to say that if evidence were to emerge that international oil flows could be this strongly affected over the next decade, substantial pressures might be exerted on the economic behaviour of the oil exporting countries. This might be so even if the above estimates of either the U.S. or the West European supply alternative, or perhaps both, are somewhat on the high side.

It is well known that Middle East and North African oil is sold at a large and growing margin over its real costs of production. Under such circumstances, if competition among exporters were for any reason to erupt – as some Western observers think may follow in the wake of greatly increased amounts of 'country-owned' oil entering world markets – these countries would be able to substantially cut their greatly inflated prices if they wanted to carve out a larger share of world markets for themselves. Today, however, there seems little likelihood that competition among the exporting countries will ever return.

But consider that while the real incremental production cost may constitute the theoretical floor to Middle Eastern oil prices, the ceiling to these prices will be determined by conventional or non-conventional energy sources available elsewhere in the world in ample amounts. We have seen how both incremental U.S. crude and shale oil are estimated at real long-term supply costs probably no higher than $38/m³ in 1972 dollars. With successive increases in Persian Gulf posted prices (the most recent in October, 1973), it will not take many additional increases for the delivered price to approach the ceiling set by the long-run cost of Western Hemisphere alternatives. Conditions thus seem to be developing which could, in time, reintroduce a note of economic reality in the pricing of crude from the Middle East.

This is a potential but not inexorable scenario, which will in any case take time to have visible effect. It does seem to offer one tangible opportunity to gradually provide a viable counter force, though not a hostile one, in international oil markets. We believe that the most pressing policy need is for the United States to embark on a coherent strategy greatly to expand its domestic productive capabilities. Such a strategy – designed to reduce excessive external dependence – seems now to be gathering increasingly widespread support in the United States, the more so in the wake of the latest developments in the Middle East. This is a policy which is clearly not a reversion to narrow economic nationalism. Indeed, it is an internationally forward-looking policy, in that it promises to blunt the enormous economic advantage exercised by the major exporters over countries – both developed and less developed – which inevitably will remain measurably dependent on substantial amounts of imported energy for many years to come. It would also make less likely a fiercely competitive scramble among buyers seeking to assure themselves of supply adequacy, with all the destabilizing international political and economic effects that this would imply. It promises, finally, to reduce the scope for international monetary disruption which a continued unilateral advantage in international oil bargaining would otherwise confer. The lines of development we have indicated offer hope that the instability which presently characterizes world oil flows, and which is bound to be with us for some time into the future, is none the less a course from which the world can yet break loose.

Phil. Trans. R. Soc. Lond. A. **276**, 431–438 (1974)
Printed in Great Britain

CONVENTIONAL PRIMARY ENERGY RESERVES: REVIEW AND DISCOVERY POTENTIAL (WORLD-WIDE)

General problems of collecting and understanding world energy data

By D. C. Ion

U.K. Member, Consultative Panel, World Energy Conference 1974 Resources Survey

This paper attempts to improve the understanding of the available world energy supply data, which are being imperfectly interpreted and increasingly used.

There are basic differences in reserves concepts and definitions between the primary energy resources. There is insufficient knowledge to justify the apparent precision of original measurements of most energy resources. The collection of energy data is in many diverse hands, but of paramount importance is collection close to source, because original estimates are progressively amended by opinions which are difficult to ascertain and assess. The factors affecting energy demand are not yet sufficiently understood to take full advantage of the potential of mathematical models. Cooperation between the energy industries must develop.

Experience in preparation for the 1974 Survey of Energy Resources by the World Energy Conference illustrates these points of which an understanding is necessary for wise consideration of world energy problems.

Concepts and definitions

The most recent attempt to define energy resources in such a way that the data are collectable and comparable is that by the Consultative Panel for the 1974 Survey of World Energy Resources for the World Energy Conference organization. This is the latest in their survey series, with that of 1968 being the last previous one. The survey should also satisfy the requirement for such a survey as recommended to the United Nations by the Stockholm Conference on the Environment, 1972.

The definitions agreed are a compromise between differing reserve concepts, and between what some planners, economists and statisticians think is required and what is practical when acknowledging that the precision of the data normally improves with the age of development of a resource and of the country in which it is found.

Solid fuels

Solid fuels include high- and low-ranking coals, peat and non-conventional fuels (wood and agricultural and domestic waste). These latter, though a small and declining proportion of world energy resources, can be important locally. Most available information is on the coals. The broad categories of anthracite, bituminous, sub-bituminous, brown coal, lignite and peat are being used. National standards vary widely and it is not always possible to define ranks by explicit reference to ash, carbon content, etc.

The widespread, general concept behind reserves data as understood by the coal industry was shown in two of the definitions used in the 1968 survey:

'Measured reserves...shall mean the total amounts...occurring within the limits hereinafter prescribed and with respect to which there exist reliable data and thickness of seams...the respective limits...shall be:

'Coals: Seams containing not less than 30 centimetres of coal and not more than 1200 metres below the surface...'

'Percentage Economically Recoverable shall mean the proportion...that is considered of economic value.'

For the 1974 survey the definitions have been changed to be more comparable with those of the petroleum industry, so recognizing four very important concepts:

(*a*) While it is important to know the total amount in place, it is more important to know what can be recovered under current operating and economic conditions.

(*b*) The thickness and depth of seams from which coal may be economically mined by current methods varies from place to place.

(*c*) The only coal which has a comparable 'economic value' is that which can be economically mined. The parameters required to judge the economic value of coals through exploitation by gasification are not yet sufficiently known to warrant inclusion in this survey.

(*d*) Changing conditions can highlight different characteristics, e.g. the request for the sulphur content of coals, (as well as for petroleum), reflects current environmental interest.

Crude oil

The 1968 survey moved its petroleum definitions from the earlier coal-type concept towards oil-industry practice and in the 1974 survey the definition of proved recoverable reserves is: '...the estimated quantities of crude oil (or liquid hydrocarbons) remaining in the ground which geological and engineering information indicate, with reasonable certainty, to be recovered in the future from known oil reservoirs based on the present state of technology and under existing economic conditions.'

This definition is similar to that of the American Petroleum Institute (A.P.I. 1969), which, however, has an important rider: 'Both drilled and undrilled acreage is considered in the estimates of proved reserves. However, the undrilled proved reserves are limited to those drilling units immediately adjacent to the developed area...'

This rider is important for three reasons. (*a*) In conditions where drilling units are small and geology complex, as in U.S.A. and Canada, this results in a very conservative estimate. In areas like the Middle East, 'drilling units' are different because the operating conditions, mineral rights laws and simple geology allow more scientific and less bureaucratic location of wells; hence greater undrilled reserves can be and are included. (*b*) The inhibition is particularly applicable, in areas like the U.S.A., to estimates in the early development of fields. (*c*) Annual revisions and additions, therefore, are much more important in areas of U.S.A. type than of Middle East type. This emphasizes the need for the constant review for both types of reserves, the one because many small additions may give large aggregates and the other because single wells can indicate large additions.

However, all countries show slight variations on the A.P.I. system, even if they subscribe to it in general, while some have very different systems.

The U.S.S.R., for instance, classify their reserves into six categories, with a system of reporting and certification, designed to ensure the application of exploration effort appropriate to the rank of prospect. There is no exact equivalent of the American type 'proved reserves' category. Furthermore, petroleum reserves are subject to the States Secrets Act, 1947 (Campbell 1968).

Natural gas and natural gas liquids

Natural gas occurs as associated or non-associated with crude oil. This raises reserves concept problems.

Associated gas is only recoverable as the crude oil is produced. Non-associated gas is a resource recoverable without reliance on any other resource. Hence a simple summation of estimated reserves would not indicate production potential. It is important to have data on both types. Furthermore, in some circumstances it may be possible to produce wet, associated gas-cap gas, strip it of its condensate and return the dry gas to the reservoir.

Another problem is related to the commitment of natural gas reserves to long-term contracts. Reporting of reserves in U.S.A. and Canada is coordinated by the A.P.I., the American Gas Association and the Canadian Petroleum Association. However, since 1966, Canadian reserves have been reported on a marketable and not recoverable basis. The difference in total in 1965 was some 115×10^9 m³ (C.P.A. 1969).

Natural gas liquids are recorded in the U.S.A. and Canada and their proved recoverable reserves indicate the estimated quantity which may be extracted by existing or planned plant in the process of producing the natural gas recoverable in proved areas. Elsewhere the natural-gas liquids reserves, and also production, will be included in the crude oil figures as liquid hydrocarbons.

Bituminous sands and oil shales

In the restrictive sense of the definition of 'Crude oil proved recoverable reserves' the only reserves from bituminous sands would be those recoverable by the existing plant at Mildred Lake in the Athabasca oil sands area of Alberta, Canada. The Canadian Petroleum Association estimate (C.P.A. 1969), was some 10^9 m³ (6.334×10^9 barrels), but they state: 'These estimates in no way detract from published estimates of approximately 300000 million barrels (48×10^9 m³) which are estimated to be recoverable from the Athabasca type oil sands by mining and thermal processes.'

The 1974 Survey is seeking information similar to the 48×10^9 m³ estimate and no mention is made of the current conditions restriction. Hence the 10^9 m³ can be compared with the 1.5×10^9 m³ (9.7×10^9 barrels) of Canada's proved liquid hydrocarbon reserves (C.P.A. 1972) but the 48×10^9 m³ must be compared with whatever figure Canada provides under the heading of Additional Reserves of crude oil.

The oil sands, therefore, provide an excellent example of the need to compare like with like. The oil shale situation is similar but the only significant commercial exploitation which could provide reserves in the strict proved sense are in the U.S.S.R. and the People's Republic of China.

Hydraulic energy

For the 1974 Survey, countries have been asked to categorize hydro-electric resources into developed, under construction, proposed for eventual development and those sites which may be economically feasible. This should give useful information more appropriate to discussions such as the present than figures of 'gross theoretical capability' which are also being requested.

Indeed, for the 1974 survey a more simple definition has been adopted than that of the 1968 survey: 'the energy potentially available if all natural flows were turbined down to sea level with 100% efficiency...'. Less comprehensive estimates may be given, using atmospheric precipitation and water run-off.

Such a gross theoretical capability is indeed gross and theoretical. Such a procedure would be impossible on several grounds (amenities, transport, sewage disposal, etc.) and the figure has very limited meaning for comparative purposes. It can only be compared to the gross estimates of other resources, e.g. the average concentration of uranium in the Earth's crust is said to be about $2.9/10^6$ by mass; hence the top kilometre of the continental crust contains about 10^{12} tonnes of uranium, which, in a thermal reactor, would have a thermal value of 1.6×10^{27} J (1.5×10^{24} Btu); current world energy consumption is estimated at 2.6×10^{20} J (2.5×10^{17} Btu) (Leslie 1973).

Nuclear energy

The resources of uranium and thorium for the 1974 survey are sought in terms of their oxides recoverable from known ore deposits within ranges of recovery costs to be given and with currently proved mining and processing technology, quoting also the average grade of ore in terms of percentage by mass of the oxides (U_3O_8 or ThO_2).

This requirement is close to the definition of 'reasonably assured resources' of the Organization for Economic Cooperation and Development/Nuclear Energy Agency (O.E.C.D. 1973), being: 'Uranium which occurs in known ore deposits of such grade, quantity and configuration that it can, within the given process range, be profitably recovered with currently proven mining and processing technology. Estimates of tonnage and grade are based on specific sample data and measurements of the deposits and on knowledge of ore-body habit.'

Recovery costs for this youngest energy resource can be spelled out. O.E.C.D., through N.E.A., and, since 1966, jointly with the International Atomic Energy Agency, have attempted to assess, for the world excluding Eastern Europe, U.S.S.R. and the People's Republic of China, the uranium recoverable within different price categories. The 1973 report, (O.E.C.D. 1973, and see also Boxer, Haussermann, Cameron & Roberts 1972), consider two price categories, up to U.S. \$22/kg U_3O_8 and between U.S. \$22 and U.S. \$33/kg U_3O_8, having already discarded in 1970 the category of between \$33 and \$66/kg of earlier reports as unlikely to be worked in the forseeable future.

Ideally this type of information could be important for all energy resources. Estimation has been attempted for petroleum in the U.S.A. (N.P.C. 1972), but with present knowledge this would be an impossible task, world-wide, at this time.

Renewable energy resources

The contribution of solar and tidal energy is, as yet, insignificant in total but may be important locally. Again, it is only installed or planned capacities which have meaning. The estimate of the total solar flux to the Earth as about 7.5×10^{16} W, with maximum intensity of 8.4 J cm^{-2} min^{-1}, compared to the total of 1970 gross energy use in the U.S.A. of 7×10^{19} J (Ubbelohde 1973), is academically interesting but of very little real value.

However, as highlighted at the U.N.E.S.C.O. World Solar Energy Congress, Paris, July 1973, research into the possibilities of solar energy is being stepped up in the U.S.A., the U.S.S.R., Israel, France and Germany, and perhaps in the 1986 survey some solar energy figures will be significant.

MEASUREMENT

Coal reserves estimates, with seams measured in centimetres, have an accuracy which is more apparent than real. The National Coal Board in Britain, the best geologically mapped country in the world, found it necessary to mount, early in 1973, a borehole drilling programme, costing £3M over two years, to locate precisely the position of potentially recoverable coal and determine its quality and best extraction methods (Broadbent 1973). In July the existence of large new deposits in Northeast England was publicized.

The 1968 survey gave the measured reserves of the U.K. as 12.2×10^9 t, with a note that 'the reserves given are those within the colliery takes and those in the immediate vicinity which are likely to be economically workable'. For the 1974 survey the National Coal Board has provided a figure for 'Known Reserves, Quantity Economically Recoverable', of 3.87×10^9 t, but with the minimum seam thickness of 61 cm (cf. 1968 stated criterion of 30 cm) and maximum depth of 1220 m (close to the 1968 criterion of 1200 m).

These British examples illustrate the problems of measurement, the critical character of the parameters and, possibly, in the latter case, the importance of the realistic application of the economic parameter.

The routine measurements taken in oil and gas fields, after some production experience, normally give sound estimates of the oil-in-place and the proven reserves, though estimates from single discovery wells can be misleading. However, the five common methods for estimating the most speculative category of possible reserves all have major disadvantages (Bakhirov & Ovanessov 1971). The shape of our knowledge of petroleum is that of an inverted pyramid – the farther up from discovery to utilization, the more we know. Yet figures for ultimate oil-in-place or ultimate recoverable reserves are often discussed with surprising arrogance. Current ignorance should engender humility; our successors will know more and find and produce more petroleum than we can estimate at this time. This does not mean that estimates should not continue to be made in the speculative areas. They can be stimulating exercises and of value as long as there is full appreciation that, being based on current knowledge, they have their limitations.

The additional resources of oil sands and oil shales are probably as well 'measured' as those of coal, but when one notes that the 1968 survey world total of coal reserves, which is a figure frequently quoted, included a figure of 1×10^{12} t for the People's Republic of China based on a 1913 estimate, there is obvious need for caution.

In the nuclear field there is the global type of gross average measurement as quoted by Leslie (1973) or calculated by Brinck (1971). The latter gives 80×10^6 t of uranium as 'possible' uranium reserves from which 'reasonably assured' reserves exploitable at prices up to \$13/kg U_3O_8 could be developed. Such estimates have a very different element of measurement than that involved in the estimates of high-grade ore reserves of the world, excluding the U.S.S.R., etc., as compiled by N.E.A./I.A.E.A., of 1.126×10^6 t of 'Reasonably Assured Resources (Reserves)' available at a price of less than U.S. \$22/kg U_3O_8.

The disparity between gross theoretical water-power capability and installed capacity has been mentioned, but at least topography and rainfall are better-known world-wide than mineral geology. In the renewable-resources field one might note that, currently, the measurement of the potential of undrilled geo-thermal hot-spring areas is virtually impossible because of the random distribution of active channels.

[29]

COLLECTION

There is no uniformity in the organizations responsible for collecting energy data within and between industries nor within and between countries. Hence to the problems concerned with the data is added that of collection.

The system adopted by the World Energy Conference is of a Consultative Panel, of people versed in the different resources from a spectrum of countries, formulating a questionnaire; a working group of the U.S. National Committee (as hosts for the 1974 Conference), sending the questionnaires to all National Committees, with a request for a nominee with whom to correspond; the working group collating the replies; the Consultative Panel review the collation; the working group finalize and the Conference publish. If in each country a small group of resource specialists assists the nominated correspondent, then this system has the two major virtues that there is the maximum use of local knowledge and, most importantly, there should be the maximum chance of collection as close to the source as possible.

INTERPRETATION AND USE

The raw material for energy data is produced by operators at colliery, field or plant and their immediate management. Thereafter, up each chain of command, fewer data but more opinion is added at each link; the original basic facts and assumptions are gradually lost and the figures averaged up or down. Despite painstaking effort to ensure that concepts and definitions are compatible, the data collected will, inevitably, have been amended by opinion.

The users of world energy data, economists, planners and managers in industries and governments, analysts and journalists of all types, students and academics, are so many and varied that it is not possible to warn all of the pitfalls on their individual roads, but some general points can be made:

(a) The effective size of an energy unit varies according to location, time and ownership (Ion 1970), the method of conversion (there is estimated to be a 30:1 difference in energy value of a unit of uranium whether used in a fast reactor or in a thermal reactor), and end-use (the effective size of a barrel of crude oil in 1900 was that of its kerosine content).

(b) Comparison of energy forms by conversion to thermal value can be misleading, for it discounts non-energy uses and process beneficiation – even in the petroleum family combining reserves of crude oil, natural gas and natural-gas liquid can be difficult, whether on heat content or price/value (Lovejoy & Homan 1965).

(c) The smooth curve of annual world crude oil productions masks: (1) differences in quality of constituent crudes and changes in quality demand, (2) fluctuations, declines and growths in individual country productions, (3) the influence of regulatory bodies, (4) the effects of price and cost changes in any or all sectors of the industry, (5) the influence of competing energy supplies and the changing pattern of end-use, (6) the sharp increases in prospective areas when knowledge and technological capabilities move together (Ion 1971).

Hence the production figures as well as the reserves figures in the often quoted reserves: production ratio must be treated with caution.

(d) Increased financial incentives rarely lead to immediate new discoveries. Explorers find resources, not accountants. Inhibition of risk-taking for lengthy periods breeds cautious managers, not gamblers (Ion 1973).

The mass media can provide many examples of mis-use through over-simplification. Discovery data are very vulnerable (Ion 1970), but so are reserves through confusion of categories (see references to Iran's reserves in *The Times* 1973 *a, b*). Such points form part of the programming of the 'mental models' of experienced users of energy data, but are difficult to program into mathematical models, which are increasingly used in forecasting and do have two major advantages:

(*a*) Records can be kept of all assumptions and facts used, whereas the detailed programs of 'mental models' are not retrievable, often not even by the 'operator', and his expensive 'machine', is lost when moved from the job.

(*b*) Mathematical models can react so quickly to amended facts and factors that the essential continuous review is possible. Only the 'published' spot checks of forecasts should be out of date, never the working papers.

On the demand side of forecasting there is an impressive list of areas of insufficient knowledge in basic assumptions (Stelzer 1974), including the societal cost of environmental conservation, and much research is required before technological advances and behavioural patterns can be incorporated into models with sufficient precision for reliability (Rothkopf & deVries 1973).

CONCLUSION

Consideration of the general problems of collecting and understanding the 'facts' of world energy resource supply, with a brief mention of demand forecasting, does not permit a forecast of the changing pattern of supply. However, there is, undoubtedly, a growing realization within the energy industries of their interdependence, their need to understand each other's strengths and weaknesses, and ensure that, by cooperation, all resources are used in the optimum manner.

Optimization of utilization will require more information about the energy resources than at present collected and published, for it will involve dovetailed projects as well as discrete ones, in order to eliminate waste and provide the greatest practical efficiency.

Such cooperation, best accomplished at all levels of industries and governments, should crystallize what data is essential and facilitate its provision and collection, as well as ensuring greater mutual understanding.

The author wishes to acknowledge the help of those from all the energy fields who commented on the first draft.

REFERENCES (Ion)

A.P.I. 1969 *Standard definitions for petroleum statistics*. Tech. Rep. no. 1, 1st ed., Stats. Dept., Washington, D.C., U.S.A.: American Petroleum Institute.
Bakhirov, A. & Ovanessov, G. 1971 *Proc. 8th Wld Pet. Con.* **2**, 315–325.
Boxer, L. W., Haussermann, W., Cameron, J. & Roberts, J. T. 1972 (Peaceful uses atomic energy) Uranium resources, production and demand. *Proc. 4th Int. Conf.* **8**, 3–22. Geneva.
Brinck, Johan 1971 M.I.M.I.C., *Eurospectra* **10**, 46–56.
Broadbent, D. H. 1973 Prospects for coal. Inst. Pet. Mtg, June 1973. (Pre-print.†)
Campbell, R. W. 1968 *The economics of Soviet oil and gas*. Baltimore: J. Hopkins.
C.P.A. 1969, 1972 *Canadian Petroleum Association statistical yearbooks*. Calgary, Alberta: C.P.A.
Ion, D. C. 1970 Communication problems in reserve concepts and in environmental control. *Proc. S.W. Legal Foundation, Explor. & Econ. of Pet. Ind.* **8**, 31–57.
Ion, D. C. 1971 Arctic oil and the world: one perspective. *Am. Ass. Pet. Geol., Arctic Symposium*. Tulsa. A.A.P.G.
Ion, D. C. 1973 Planning for the future. Inst. Pet. Mtg, June 1973. (Pre-print.†)
Leslie, D. C. 1973 Nuclear energy 1. Inst. Pet. Mtg, June 1973. (Pre-print.†)

Lovejoy, W. F. & Homan, P. T. 1965 *Methods of estimating reserves, etc.* Baltimore: J. Hopkins.

N.P.C. 1972 *United States energy outlook: a summary report.* Washington, D.C.: National Petroleum Council.

O.E.C.D. 1973 *Uranium resources, production and demand* (N.E.A. & I.A.E.A.). Paris: O.E.C.D.

Rothkopf, M. H. & deVries, H. 1973 Modelling future energy supply. Inst. Pet. Mtg, June 1973. (Pre-print.†)

Stelzer, I. M. 1974 Advance copy of Position Paper for Division I, Ninth World Energy Conf., Sept., 1974, Detroit.

Times 1973 Times newspaper, London. (*a*) 26 June (p. 22). (*b*) 16 July (p. 15).

Ubbelohde, A. R. 1973 Alternative sources of energy. Inst. Pet. Mtg, June 1973. (Pre-print.†)

† These papers are to be published, December 1973, as *Proc.* 1973 *Inst. Pet. Summer Mtg*, 'Energy – from surplus to scarcity'. Applied Science Publishers.

Discussion

PROFESSOR D. C. LESLIE (*Queen Mary College*): *The extent of the reserves of nuclear energy*

Mr Ion has stressed the difference between proven and estimated reserves, and has suggested that one should not assume too readily that the estimated reserves will necessarily become available. One can only agree with this wise note of caution, but I would like to stress the great difference in this respect between fossil and nuclear energy reserves. When dealing with fossil reserves one encounters statements such as 'the proven reserves amount to 20 years consumption, the estimated reserves to 50 years'. When dealing with nuclear resources the proven reserves may well be of the same order as the fossil reserves. However, in the nuclear case the estimated reserves are good for millions of years. Therefore, even if only a small fraction of the estimated reserves can be brought into play, none the less the inclusion of this small fraction alters the overall picture beyond all recognition. This is obviously not true in the fossil case.

Furthermore Mr Ion has emphasized that we do not really know where to find the estimated fossil resources. This is not nearly so true of the estimated nuclear resources. Granites contain around 4 parts/10^6 of uranium and we certainly know where to find granite in very large amounts. Comparison with open-cast mining of copper suggests that it might be possible to produce natural uranium from granite at less than £5/g. If burnt in a fast reactor, which was able to fission half of all the uranium atoms, the contribution of the purchase of uranium at this price to the overall cost of electricity would be about 0.1 p/kW h (36 p/GJ). Compared with the current wholesale price of electricity of (say) 0.5 p/kW h (180 p/GJ) it is clear that cost of this uranium is by no means insupportable.

In order to sustain the current world total energy consumption in this way we should have to mine around 2×10^9 t of granite each year. This is comparable with the present rate of extraction of coal and poses a serious but not insuperable environmental problem. Moreover there will be no need to mine granite for a very long while yet. There are very large uranium deposits at concentrations of around 50 parts/10^6: these would be both cheaper and less messy to extract.

Phil. Trans. R. Soc. Lond. A. **276**, 439–452 (1974)

Printed in Great Britain

World coal resources and their future potential

By G. Armstrong

Mining Department, National Coal Board, London, SW1X 7AE

Despite the uncertainties which arise from drawing inferences from the published figures of world coal resources which are based on a variety of assessment procedures, there can be no doubt that coal is the world's most abundant fossil fuel, with total reserves of the order of 9×10^{12} tonnes.

Industry, however, can only be interested in those reserves which are exploitable now or seem likely to be in the future. As time passes, the proportion of recoverable reserves varies and is controlled by advances in technology, changes in the economic circumstances and the level of exploration.

There is a continuing need to balance reserve availability and productive capacity against present and future demand. The locating of economically recoverable coal reserves takes time, and so does the build-up of the necessary market requirements (e.g. construction of new power stations). When there is a need to increase or replace productive capacity, it is important to plan well in advance in the exploration, production and marketing phases.

Timely investment in world coal exploration and productive capacity appears essential in view of possible shortages of other fuels.

1. Introduction

Demand for energy in one form or another will always be with us and the balance of utilization of the world's basic sources will depend on their availability at currently competitive prices. In regard to coal, there are a number of fundamental controls which affect its economic availability. These are partly geological, with which must be linked the engineering resources appropriate to a particular geological environment, and partly economic and logistic. The economic circumstances are mainly determined by the price levels of competitive fuels, and logistics by the location of reserves in relation to markets and feasibility of cheap transport.

It is important to note that, apart from the basic geology of the deposits, all the remaining factors, including the degree of knowledge of geology, are variable with time. Engineering efficiency can improve, albeit generally slowly, with fresh technical innovations, as can transport; while the economic climate is probably the greatest variable and is likely to change rapidly and unexpectedly.

Because of all these variables, estimates of reserves, particularly those thought to be economically recoverable, can vary from time to time even though the basic geological environment is fixed and the gross total of coal reserves changes hardly at all.

2. Geological factors affecting assessments of reserves

The principal coal reserves of the world are geographically widely distributed and spread over a very broad band of geological time ranging from the Carboniferous to the Tertiary. The bulk of the economically important reserves, however, are in the Carboniferous of Europe and eastern North America and the Permian of the Southern Hemisphere and Asia, while the bulk of the western reserves of the U.S. and Canada are of Cretaceous age. In addition to its widespread geographical distribution, and its extensive spread of age of origin, coal occurs in a variety of geological environments depending on the original conditions of deposition and the effect of subsequent metamorphosis, mainly arising from temperature changes generated by compressive earth forces.

The substance we call coal, therefore, is sometimes lignite, sometimes bituminous coal, or anthracite or an intermediate type. Each type can contain, depending on its origin, variable amounts of dirt, sulphur and other constituents, all of which can affect the value of the coal and hence the proportion of the gross reserves considered at any one time to be economically workable. The geological conditions which largely control the ease with which the coal is mined are, however, the basic controlling factors in determining reserve assessments.

The sedimentary environment of the coal seam demonstrates the original mode of origin and is of critical importance. Different regional rates of variation may be exhibited by different sedimentary phenomena which include the coal seam structure itself, i.e. incidence of splitting and the thickness and disposition of bands of dirt and inferior coal, together with its overall thickness and physical and chemical properties. The sedimentary environment has also given rise to such vital factors as the composition and physical properties of the roof and floor strata, the occurrence of washouts and associated compactional disturbances.

Earth forces subsequent to the original deposition of the coal create the tectonic environment in which the coal seams are found today, and this is a particularly important element in assessing the workable reserves in many parts of the world. The rates of regional change exhibited by the different tectonic phenomena, which include gradient, faulting, jointing and cleat, need to be considered in conjunction with the sedimentary factors, but probably the most important single item is the actual intensity of faulting which can seriously retard the continuous cycle of mining operations that is so necessary for economical working. The regional (lateral) and vertical rates of change of the various geological factors affect the economic workability of an area of reserves and determine how closely spaced the points of geological information must be for the allocation of those reserves into the recoverable category. In Britain, an area where the information points are over 3.2 km apart would normally be excluded from reserves in this category.

3. ENGINEERING FACTORS AFFECTING ASSESSMENTS OF RESERVES

The economically recoverable reserves in a coalfield depend to a very large extent on the capability of the mining systems in use to cope with the difficulties presented by the particular geological environment. In many parts of the world, in recent years, mining systems have been evolved which help minimize the effect of some types of adverse geology and ensure high levels of productivity. In Western Europe, where mines are now virtually 100 % mechanized, this means that coal is cut on the face mechanically, is conveyed automatically to the shafts, hauled to the surface, cleaned and beneficiated in modern plants and loaded into wagons for transport direct to the markets.

The very sophistication of these systems – designed to produce coal at minimum cost – inevitably rules out the working of large areas of coal where the geological conditions are unsuitable. In the past, reserves of marginal value were often included when estimating the potential of a colliery or a coalfield, in the belief that improved technology would enable this coal to be worked economically in the future. In fact, to date, the reverse has been true, in that the modern rather inflexible mining systems require coal seams to be in a particularly favourable geological environment in order to support intense and continuous output which is needed for economic production. In old coalfields reserves lying in these favourable conditions are becoming increasingly difficult to locate, so that there is a real need to evolve more flexible mining systems capable of dealing with a greater range of geological conditions, and for mining layouts specifically planned to minimize the effects of stress and instability caused by depth and previous workings. Here indeed is a rich field for research which could well lead to some increases in economically recoverable reserves in many of the world's developed coalfields.

4. ECONOMIC FACTORS AFFECTING ESTIMATES OF COAL RESERVES

(a) The nature of coal deposits and the relative ease with which they are found

There are physical differences between deposits of coal and those of competing energy minerals which render the processes of recovery the most cost-significant to coal industries throughout the world. Coalfields usually consist of a multiplicity of flattish sheets of coal, all set in a very distinctive suite of enclosing rocks and generally of huge areal extent (although the individual seams form only a small percentage of the total ground within the coalfield). Thus when their edges crop out at the surface they are an unmissable target to the geologist and mining engineer. When concealed, one exploratory hole is often enough to 'prove', in the sense of test or sample, the presence of a concealed coalfield. In contrast, major new oil or gas fields must, for their existence, always be sealed beneath other strata and can occur as pockets of small areal extent in almost any strata that are sufficiently porous.

Because a concealed coalfield competes with its exposed neighbours for exploitation, it is depth that counts initially, almost more than anything else. Unlike gas or oil, coal does not flow out of its enclosing rocks towards the discovery hole under the huge natural pressures in overlying strata, so we have to go down and get it. Because this is so, and because in all of the world's coalfields except one or two temperatures increase with depth, depth of extraction is limited to the temperatures that we can work in without expensive cooling processes. In world terms the deepest discovery hole for any new mineable concealed coalfield is never more than a thousand or so metres. By contrast, such relatively shallow depths rarely provide sufficient pressure from the overlying rock to squeeze petroleum out of the strata in economic quantities. Thus search depths for target bodies of petroleum are measured in kilometres. These bodies are also relatively small compared with coal deposits, and not in any unique suite of associated strata. Hence greater weight is given to risk capital expenditure on finding new fields of petroleum than of coal. Only in regard to petroleum is exploration the most economically sensitive link in the chain between the energy mineral as it occurs in the ground and as it is finally utilized.

For the many reasons given above I suggest that, unlike petroleum deposits, the location of the vast majority of the world's coalfields, in terms of gross tonnage of coal, have already been discovered. Since, as will be shown, we are nowhere near having exhausted the coalfields already discovered in the world, there is relatively little potential for greater world-wide expenditure on the search for completely new coalfields.

(b) The nature of coal deposits and the relative difficulty with which they are recovered

In the petroleum field the mineral flows through the reservoir rocks towards the access points in variable proportions through routes which do not have to be known in advance and are not discovered in any detail during extraction. In deep mining for coal each route into the deposit

[35]

and precise location where continuity of extraction can be obtained is, at every stage in life of the field, a matter for management decision. The position of each minor geological change in the relatively thin coal seams has either to be known in advance or is expensively encountered after the decision has been made.

The significance of having to move into the deposit – instead of only having to drill a hole for earth pressure to push it out to the surface – is that in coal mining decisions can be made concerning precisely which portions and how much in each portion shall be extracted.

(c) The economic influence of depth on recovery of the world's coal reserves

Only a limited proportion of coal in the world's coalfields lies at opencast depths, of less than a hundred or so metres. These reserves can only be a small percentage of the world's total, since most of the mineable coalfields lie between the surface and a thousand or so metres. The capital expenditure on facilities for making access to opencast coal, however, is both highly productive in manpower terms and, unlike that for making access to deep-mine coal is, to a large degree, transferable to other sites. Thus the history of most of the recently discovered coalfields is mainly one of recovery by the opencast method. Just under half of the coal from the fields of the United States and more than half of the U.S.S.R. output is still produced by this inherently cheap method, and soon 50 % of Australian output will be opencast as well. In Britain we are unable to mine a similar proportion because such coal in our fields was extracted by less-efficient shallow mining, before opencast methods permitted mining more than a few metres below the surface.

If mining is to recover more than the small fraction of the world's coal resources which are available to opencast, it must obviously be by deep-mining methods. However, the expenditure on making initial access for deep mining is enormous, and most of the capital equipment which has to be provided is not transferable but thereafter is only serviceable out of revenue. Historically, once provided, these access points generally last four or five times as long as those provided for access to oil and gas fields. The long delays before an earning position is reached offset the potential they still offer in their hundredth year, which is concealed by the compound interest rates used in cash-flow accounting.

(d) The effect of geological conditions on the economy of existing deep mines

Let us, for the moment, defer consideration of the prospect of new deep mines and restrict ourselves to such of the world's deep mines as are currently in operation. One fact of coal mining not always clearly appreciated is the extent to which the variable nature of coal deposits affects the economics of running a mine. Nearly all coalfields show a varying thickness of seam and a degree of local geological change or disturbance.

In any deep mine in the U.K., the greatest financial benefit can generally be attributed to the most productive of its longwall faces. In America and Australia where longwall faces are few, it may be attributed to the most productive of the mechanized pillar or shortwall districts in the mine.

In deep mining, day-to-day costs are largely costs per unit of geologically undisturbed area worked, while for a given quality, the revenue at 'the pithead' is obviously per unit volume worked and raised. (This is why in the majority of the world's existing mines the areas in which the seams are thicker, and of good quality and reasonable accessibility, have already been worked in the less disturbed districts of the mine.)

[36]

Any local change of geological circumstance which worsens the thickness, or the continuity of the coal, or the face conditions being enjoyed by the best unit in the mine when the time comes to replace it, strikes directly at the profitability, as there is no significant change in overheads. This is because wages and interest charges on the capital equipment on all the faces, and the logistic system behind them are the major items of cost. The deployment of effort and investment among units within a mine, and among mines within a company or national mining industry, is such that those temporarily enjoying the best conditions commonly support a large number of 'average' or 'worse than average' units. In a major company or national context loss of one of these 'best conditions units' therefore has a highly geared numeric effect on the total number of access points which can be maintained in production. It is this aspect of an extractive industry, quite naturally first taking the easiest coal to win, which may be unfamiliar.

In world terms, when there is a period of cheaper energy, quite properly no one wants to finance large coal stocks for long periods. Thus it is nearly always a number of the less economic mines which have to close. Similarly, within the remaining mines, it is the least economic seams in individual mines which have to be written off, when the decision on which seam or district to work next has to be made. Thus, over the last decade, we have a massive reduction in the number of access points in the coalfields of the European coal industry. Let us now examine the manner in which the process of recovery of the world's coal resources is affected by alternative sources of energy, by reference to the case history of Britain.

(e) The influence of energy economics on recoverable reserves – an example from Britain

In a previous energy crisis, precipitated by a long history of massive growth, the Royal Commission of 1871 lamented that there was a considerable loss of reserves where 'the whole or part of a bed adjoining or lying near to that which is being worked is left behind, where it becomes crushed and unworkable hereafter, because it is of a kind which is not wanted at the time, or if it were worked the cost per ton of coal got would be increased'. This is still true today.

Since about 1900 each successive choice of what coal to work next has had to be taken in competition with petroleum as a significant alternative to coal as a source of energy. It was a case of margin against margin, to hold coal's dominant position against oil in the British energy economy throughout the first half of the century.

What happened took place in spite of Dr Boverton Redwood's evidence before the next Royal Commission required to advise on the future of coal, when in 1901 he reassured them that for 'anything like general employment (of oil) I cannot see where we are to get the supplies'. (World coal production was 790×10^6 t and oil production only 22×10^6 t in 1901.) The Commission rested on expert assurance that if the future of the country required energy it could only come from coal. Thus, whatever portion of the resources was physically workable even if currently too thin to be economic, would at some time or other in the future necessarily sustain the life of the industry as long as the country required energy. The coal resources, down to seams of about 30 cm in thickness, and under 1220 m deep, were therefore nearly all to be included as 'reserves' – in the view of those required to foresee the climate of energy economics at that time.

(f) The source of confusion in estimating the life of coal reserves

Most of the confusion concerning what should constitute the reserves on which an estimate of the life of a deep-mining industry should be based arises from the fact that it is possible to probe into the future in terms of the presence and condition of the mineral to be extracted in future years. In this sense, extraction by movements through the deposit in future space are equivalent to increments of time, as far as supply is concerned. Because we drill and sample the condition of the mineral ahead, it is possible to estimate the total volumes present in a field which has been thus sampled, or in this sense 'proved'.

Once the physical limitations on what constitutes reserves have been authoritatively agreed, assessments are a matter for sophisticated measurement and calculation. The most relevant figure, with regard to total rate of demand, is the latest average annual production. It is all too simple to divide the one figure of annual demand into the other figure of total reserves in the fields concerned – to obtain an estimate of the life of the world's mineral-energy supplies. Since, however, division of one figure by another is multiple subtraction, each successive year's demand subtracted from the preceding number of tonnes of reserves has a different effect on the remainder, depending on what coal it has been necessary to select and what has been rendered unworkable or too expensive in the process of selection – in other words (as the Commission of 1871 pointed out) on what has happened in the changing climate of energy economics in the previous years. To examine this effect further we must return to the case-history of Britain, remembering that, unlike oilfields, coalfields when discovered in the past usually lasted for hundreds of years.

(g) The trend in Britain in successive estimates of economically recoverable reserves

The Royal Commission of 1901–5, referred to above, took evidence of what coal seams could be sensibly considered as likely to be capable of sustaining the life of the industry, and of what losses occurred during extraction. Armed with this evidence, their geologists and surveyors then added up the quantity of coal amongst the total resources which lay within the limits imposed, distinguishing such tonnages as were uncertain due to an inadequate degree of physical proving. They did not feel that they should be constrained by any considerations of a time limit on extraction. As we have seen, they thought that there was no real alternative fuel to coal. They obtained a British coal reserves figure of some 144×10^9 t (equivalent to a reserve of between 600 and 700×10^9 barrels of oil, or *ca.* 4.15×10^{21} J, depending on the conversion factor used).

There was renewed apprehension about the sufficiency of coal supplies during the Second World War, and the Department of Scientific and Industrial Research again took similar evidence. However, in the light of the circumstances of the time, they decided that it was prudent to restrict their inquiry to reserves obtainable only over the next 100 years. After due measurement they arrived at the figure of 20.8×10^9 t. A further authoritative assessment of the reserves relevant to the coal industry in Britain was made, this time by the National Coal Board for National Plan purposes, in the late 1950s and early 1960s. In the circumstances of the time (rapidly falling world oil prices) it was considered prudent to consider only coal which was accessible to existing or developing mines, as no capital seemed likely to be forthcoming for any new mines in the foreseeable future. This new assessment revealed some 16×10^9 t of reserves thought likely to be capable of sustaining the industry.

(h)   The depletion rate concept of currently accessible and economically workable reserves

The trend over these widely spaced intervals of assessment represents a write-off rate, of reserves successively proving likely to be uneconomic to extract, averaging more than 2×10^9 t per year; while actual annual extraction was only a tenth to a twentieth of this amount. The reserves figures were still comfortingly large in absolute terms, but the National Coal Board decided to update the reserves assessment annually instead of leaving it to intervals of a decade or two. The subsequent assessments confirmed that reserves which were considered workable under the national minimum productivity requirements of previous years were still falling annually as national productivity requirements increased. An assessment this year (1973) gave a national total of current workable reserves of some 3.9×10^9 t – enough for nearly 30 years at the current rate of extraction. And it must be remembered that the thousands of millions of tonnes in the virgin portions of the coalfields not yet assessed are still there.

Fortunately, over the 1960s, the last wholly human link in the chain of actions involved in getting the coal from the solid seam into a surface transport system was removed. It was achieved by simultaneously cutting and loading the coal mechanically and then mechanizing all the support and logistic equipment necessary to keep up with the enormously increased rates at which faces could advance into the coal. This final step, in a long succession of technological revolutions in the coal industry, took the usual decade to become disseminated throughout the industry. As is well known, national productivity – in both the thickest and thinnest seams – rose rapidly.

Mines, during this decade, were closed at the rate of some forty per year. The number of faces in the mines remaining open were reduced some fourfold. Output over the decade was reduced by one-quarter (being replaced by cheaper oil) but productivity increased by over 50 %. Without this technological revolution there is little doubt that British coal-mining capacity would have been far more than halved under the onslaught of cheap world oil prices during this decade. Had this happened we would have greatly increased the write-off rate of currently accessible economically workable reserves, instead of being able to hold it more or less constant.

Our concern still lies with the rate of annual write-off and loss of access to reserves which can, under the economic conditions and the technology prevailing at the time of each annual assessment, be considered as likely to be capable of sustaining our efforts in existing mines. In deep coal-mining this has proved to be a far more important factor than the rate of output. The write-off rate is of course directly related to national productivity and costs, and hence, given continued good management, also subject to the advent of fundamentally new technology.

(i)   The influence of the depletion rate of coal reserves on planning national energy economics

The steady rate of loss in workable reserves, despite our increased productivity, was due to the advent of alternative sources of energy over the last seventy years. Fortunately, during the last few years there has been some evidence that the long-run average annual rate may be flattening out. If capital is provided to make access to the best seams in the vast extensions of our existing coalfields, and if our next technological revolution can enable wages costs to rise and industry to retain the men we need, then we should, for the first time since the latter part of the nineteenth century, see an increase in the reserves of coal in Britain which can be considered as economically workable. Provided we can foresee a sufficient number of years of economic supply in the ground to meet foreseeable requirements, then we can respond by taking whatever action is necessary for extracting it at the required rate. The time-horizon with which it is necessary to view coal reserves, provided their assessment is updated annually, is about a decade ahead, as this is the time it takes to make completely new access – to replace reserves which are being depleted at the prevailing rate.

The British case-history is one of loss of access to the coal in our fields at an average rate which, if projected forward, could lead to a hypothetical zero of economically workable reserves within a few years. The danger point occurs when this zero is in sight before new access can be created to replace the loss of economically productive capacity in an extractive energy industry, under competition from temporarily cheaper sources of energy. I am sure this British experience can be duplicated in the other older coalfields of the world.

The point so little realized is that it takes nearly a decade to explore for and select the best site, sink each wholly new access point in deep mining, and bring it into full production. This is a longer response time than either finding and bringing into production a new oil or gas field, or siting and bringing on stream a new nuclear power station.

(j) The influence of response time to change on multi-fuel energy economics

Thus it is not a matter of today's relative cost of the three major sources of energy – nuclear, petroleum and coal – but the trends showing their likely costs within their response time to change. And, as we have seen, deep-mined coal has already been responding in the sector peculiar to it over the last seven decades.

It has been shown that it is valueless to add up the world's total reserves of coal and divide by the world's current production. This gives, if the British case-history has any relevance, a world life figure so large that decisions in this sector of the world's energy economy can be safely deferred to the next generation – while we turn our attention to the problems of ownership of the remaining oil. However, we were the first country to start studying this massive process of interaction in energy economics between coal and its new competitors, and we can now perceive what is involved in the process.

We are, in consequence, the first to have to engineer a more efficient response, both to the problems of increasing depth and to a world concept of investment criteria fundamentally unsuited to the huge new scale of, and the long response times inherent in, the energy economy.

Whatever the return on investment in response to more ephemeral human needs, few need reminding that the economic edifice, mainly concerned with the efficient production and distribution of an ever-growing abundance of services, goods and manufactured materials, rests on the security of the world's water, food and energy supplies. We have now a better understanding of the nature of coal reserves with regard to the action such knowledge necessitates.

The way in which these reserves are now viewed is responsive to economic change and cost, and is cast in terms of the level at which a coal industry dependent on those reserves can live (without unforseeable change in current operations). As in all things it is necessary to review and update these assessments continuously, giving emphasis to accuracy within the minimum planning horizon required for our operations. To us, as an industry, knowledge of the future beyond this horizon is unnecessary. It is up to us as a nation to make use of our knowledge of depletion trends in solving the problems which now face us – and will, later, face other owners of coalfields still enjoying a cheap labour supply or largely opencast reserves.

5. Estimates of world coal reserves

(a) Basis of estimates

There is no absolute measure of total world reserves of coal, nor in the nature of things is there ever likely to be one. Cost alone prohibits the comprehensive probing of the Earth's crust and such a programme would be of limited value as industry can only be interested in the resources that are exploitable now or seem likely to be in the foreseeable future. As time passes, the standards of exploitability keep changing as the controls change with advances in technology and changes in the economic climate. It follows that the procedures for assessing reserves will also change.

The various estimates of world coal resources that have been published are summations derived from assessments for different countries based upon differing procedures. Some of these assessments are of the total quantity of coal in place in known and inferred deposits; others are of the quantity of coal physically workable by current mining technology from these deposits; others, again, are of the quantity of coal economically recoverable from these deposits. It is therefore desirable to consider, in the light of the discussion in §§2, 3 and 4 of this paper, the degrees of accuracy or uncertainty associated with these different procedures, and the relative magnitude of these three categories of reserves.

(b) Categories of reserves and their assessment

Gross reserves

In the assessment of 'gross' reserves, that is the total amount of coal in deposits, the only coal usually omitted is that in very thin seams (say, less than 30 cm in thickness) and that in seams at great depth (say, more than 1220 m below the surface). The assessment is thus an arithmetical exercise based upon knowledge of the areal extent, contours, thickness variations and density variations of each seam, and the accuracy of the assessment is solely dependent upon the amount and reliability of the information available or obtainable on these factors. This information is obtained from actual measurements in the cases of 'known' deposits, and has to be estimated in the cases of 'inferred' deposits.

It follows that changes in estimates of gross reserves can only result from:

(i) more precise information gained from subsequent proving of the deposits;

(ii) subsequent extraction of reserves;

(iii) subsequent discovery of extensions of deposits; or of additional deposits.

Physically workable reserves

In assessing 'physically workable' or 'total available' reserves, deductions are first made from gross reserves in all areas in which, under present mining technology, a seam is considered unworkable because:

(i) it is too disturbed geologically;

(ii) it is too close to an adjacent seam, which is included in the assessment, for both to be worked;

(iii) it is too badly affected by underworking or overworking;

(iv) it is too wet, through proximity either to a natural aquifer or to adjacent flooded workings;

[41]

(v) it is required to remain unworked to support property and surface installations, or to maintain land drainage.

It is usual also, in such an assessment, to discount areas in which a seam is thought to be too poor in quality (e.g. with too high ash and/or sulphur, or with too low a ratio of coal to associated dirt) to ever become economically workable.

It is unlikely that foreseeable advances in conventional mining technology will reduce significantly the proportion of gross reserves that has to be discounted in the assessment of total available reserves. A substantial reduction seems possible only from the successful development of a completely different method of extraction (e.g. in-seam gasification or solvent extraction).

Estimates of physically workable reserves, therefore, differ essentially from estimates of gross reserves in that they are based upon additional factors which are incapable of precise measurement, except perhaps in an area where they have been well established by extensive mining experience. Also, estimates of future changes in the proportion of physically workable reserves to gross reserves are subject to a high degree of uncertainty in the longer term, as these changes would appear to depend principally upon developments of new methods of extraction.

Economically recoverable reserves

The assessment of 'economically recoverable' reserves involves attempting to estimate the proportion of the physically workable reserves that will be exploited under current trends in economic constraints. To the physically workable reserves a percentage deduction must always be applied as an allowance for the effects of geological hazards such as faults and washouts, coal that has to be left in pillars and barriers, etc. In addition, however, the assessment of economically recoverable reserves involves making judgements on proceeds and profitability, based upon estimates of the availability, deployment and cost of mining resources (manpower, machinery and materials) and the marketability and prices of the products. These estimates are subject to constant change in the light of political, social and financial developments in the country concerned and the world at large, irrespective of and in addition to any changes in knowledge of the physical disposition of the deposits, or any changes in the proportion of them considered to be physically workable. This is why, to obtain a trend in economically recoverable reserves, it is necessary to make use of a succession of recent estimates.

Estimates of economically recoverable reserves are thus subject to a very high degree of uncertainty, and also to continuous change from factors independent of the nature of the deposits.

(c) Relative magnitude of different categories of reserves

The relative magnitude of these different categories of reserves probably varies fairly widely for different deposits throughout the world, but the following two examples on a national summary basis may give some indication of the possible average worldwide relativity.

The latest assessment of gross reserves in the United States (quoted by E. H. Reichl in 3rd Robens Coal Science Lecture in October 1973) is 2.60×10^{12} t in seams to a maximum depth of 915 m, or 2.91×10^{12} t in seams to a maximum depth of 1830 m. Of these amounts less than one-eighth, some 317×10^9 t, are in 'measured and indicated' (in contrast to 'inferred') deposits, within 305 m of the surface, and can thus be considered physically workable by current United States mining practice. Economically recoverable reserves are reckoned to be contained within these deposits and are assessed at about 136×10^9 t, representing only about 5 % of the possible gross reserves of the country. About 30 % of these 136×10^9 t are recoverable by surface mining.

The latest assessment of gross reserves in Great Britain, made by the National Coal Board for the 1974 World Energy Conference, shows a total of about 163×10^9 t, of which some 99×10^9 t are in 'known' (in contrast to 'indicated and inferred') deposits. These figures relate, in general, to seams with a minimum thickness of 60 cm and within 1220 m of the surface. The physically workable coal accessible from existing or new collieries probably amounts to less than 15×10^9 t, or less than one-tenth of the possible gross reserves; while the economically recoverable coal is currently assessed at approximately 3.9×10^9 t, representing only about $2\frac{1}{2}$ % of the possible gross reserves. About 8 % of these 3.9×10^9 t are recoverable by surface mining. Additionally, however, there are some 2.5×10^6 t available to existing and planned collieries, of which a considerable proportion might become economically workable with beneficial changes in economics and/or technology.

The relative magnitude of gross, physically workable and economically recoverable reserves therefore appears to be of the same general order in the United States and Great Britain; the differences probably reflect the generally easier conditions of mining in the United States.

6. WORLD RESERVES OF ECONOMICALLY RECOVERABLE COAL

The many uncertainties inherent in estimates of world coal reserves, arising from the different bases of assessment referred to in §5 (a) of this paper, have now been considered in detail. The most recent estimate of world coal reserves published in 1968 by the World Energy Conference was over 8.8×10^{12} t, comprising 6.7×10^{12} t of bituminous coal and anthracite, and 2.1×10^{12} t of brown coal and lignite. The figures for the United States and United Kingdom were, respectively, 1.5×10^{12} t and 15.5×10^9 t. For these two countries, then, the figures published by the World Energy Conference represent respectively 10 and 4 times the current estimate of economically recoverable reserves, and agree more closely with what would be the current estimate of physically workable reserves.

If this situation is broadly true of other countries, and if it is not unreasonable to assume that 50 % of the physically workable reserves will *ultimately* be economically recovered, we arrive at a world total of 4.4×10^{12} t of economically recoverable coal. This figure is enormous, representing as it does about 1500 times the 1970 world output of about 2.9×10^9 t of coal. Assuming that world coal requirements will increase by 5 % per annum until the end of this century, world coal output in the year 2000 will have risen to over 12×10^9 t, and the total coal extracted between now and then will have amounted to almost 0.2×10^{12} t. However, sufficient reserves would still be remaining to sustain output at the level of year 2000 for over 300 years. Even if we assume that only 10 % of the estimated 8.8×10^{12} t will be economically recoverable – a most pessimistic assumption – there would still be sufficient coal to meet this assumed rate of demand to the end of the century and then to sustain output at that level for over 50 years.

It must, however, be emphasized that world reserves of coal are very unequally distributed among the continents and political blocs, as indicated by the figures at the top of the next page.

It is therefore very probable that, despite the overall abundance of world reserves of coal, severe shortages could occur at a fairly early date in particular regions or countries.

area	approximate percentage of world coal reserves
U.S.S.R.	$61\frac{1}{2}$
North America	$17\frac{1}{2}$
Asia (mainly China)	17
Europe	$2\frac{1}{2}$
Africa	$1\frac{1}{2}$
South America	$< \frac{1}{2}$
Australasia	$< \frac{1}{2}$

7. THE BALANCE BETWEEN RESERVES, PRODUCTIVE CAPACITY AND DEMAND

I have already referred to the depletion of British coal reserves. During the 1960s the declining competitive position of coal resulted in a rapid reduction in the coal reserves regarded as economic. These fell from about 16×10^9 t in 1962 to about 6×10^9 t in 1972. This process was accompanied by an accelerated rate of colliery closures on economic or exhaustion grounds, the annual number of operating collieries falling from 616 in 1962 (producing 191×10^6 t of coal) to 281 collieries (producing only 129×10^6 t of coal) in 1972/3.

Coal mining is an extractive industry with the end-point of each individual venture the exhaustion of its natural resources. In Britain, despite a massive rebuilding programme after 1947, the average age of existing collieries approaches 80 years. If there is not continuous investment to prove additional reserves and to provide replacement capacity, the industry as a whole will lose a very substantial part of its present potential by the mid-1980s, even to a point where the total output could well be of the order of only 70–80×10^6 t per annum, and this at a time when it is estimated that the demand for primary energy in the U.K. will increase by about 40 % to about 460×10^6 t of coal equivalent (13×10^{18} J) to 1985.

There is, therefore, a continuous need for all countries to balance not only the availability of economically recoverable reserves, but also the productive capacity of their coal-mining industries against present and future demand in the world as a whole. In fact, in many countries, decisions have already been reached about their forward coal policy. Probably the most important, and certainly the most impressive because of its scale, is the decision of the United States government to increase the annual output of coal from the present level of about 550×10^6 t per annum to about 1000×10^6 t by the year 1985. This expansion is to be followed by a further massive increase to an annual rate of 1400×10^9 t by the year 2000. To support this massive increase in output, the United States has, at the same time, launched a £4000M research programme covering the whole range of activities from exploration to utilization. That the United States has a coal reserve potential capable of supporting such massive increases is not really in doubt, but an increasing proportion of the output will have to be obtained from deep mines as opposed to opencast production and the locating of economically workable reserves is likely to prove slow and laborious. Important policy decisions have also been made by other major coal-producing countries, certainly in the U.S.S.R. where production, at present slightly less than that of the United States, is being greatly expanded. Plans for expansion have also been made known for China, Poland, South Africa and Australia.

At present Western Europe is a notable exception, probably because the remaining reserves, although plentiful, are generally deep and the seams are relatively thin. But most important of all, the coalfields are often situated in difficult geological environments with heavy faulting, making mining expensive and conditions difficult. From the technical point of view, therefore,

[44]

there are major problems in increasing the amount of coal mined in Western Europe or even in stabilizing output at the present level. The question which is now being debated by governments within the E.E.C. is precisely what part coal should play in the total energy economy of the Community.

In Britain, we are planning the future of the coal industry in the light of the problems I have outlined, and the world energy situation. We regard our plans as falling into three distinct chronological periods, although each impinges on and overlaps the other. For the rest of this decade up to 1980, the amount of coal we can produce must almost exclusively come from existing collieries, because of the long lead time to prove new reserves and establish new capacity. Major investment will be needed in existing pits to maximize viable production when pits have a long potential life and access to adequate reserves. The second phase, looking into the decade of the 1980s, is the period when any new pits which we sink will be producing, and we are now exploring intensively for new and productive coal reserves and already conducting a great deal of necessary immediate planning. In North Yorkshire, for example, we have already located quite shallow reserves in thick seams, amounting to about 10^9 t – sufficient to support a feasible annual output of 10×10^6 t for many years. Other parts of Britain are also being explored and it is anticipated that we shall prove sufficient reserves both in new coalfield areas and near existing collieries to support an annual output of the order of 30×10^6 t for many years, reaching full production in the mid-1980s.

From the end of the 1980s, throughout the 1990s and beyond, it is possible that we shall be doing what the United States will probably be doing a stage earlier, which is to use coal mainly as a feedstock for conversion into other suitable forms of energy. For this period, doubtless, new forms of mining or coal extraction will be required – perhaps to exploit the massive quantities of coal which lie deep beneath the North Sea out of reach of the mining methods of today.

In Britain, therefore, we can foresee a major departure from the policies pursued in the last decade or so. Whereas in that period we have been closing pits owing to exhaustion, or because of high cost, without replacing the loss with new capacity, the time has now come to bring new capacity into operation without delay. This is a big departure and will involve a very major increase in investment in the coal industry. We have demonstrated that the reserves are there waiting to be exploited. Now we have to convince the government that this is the right policy.

A policy of expanding coal output in Britain must, of course, not be taken in isolation. It should be part of an overall policy, and in Britain we have a unique opportunity of being able to develop a number of indigenous resources, on land and under the sea, in the form of gas, oil and coal. What we are aiming at here – and I am sure that this will emerge with time – is an integrated policy of forward energy planning, making full use of all the resources available, ensuring that each will be used to the maximum advantage.

8. CONCLUSIONS

Let me end by emphasizing a number of conclusions which arise out of the points made above. Taking into account only the very conservative figures of recoverable world coal reserves, there should be sufficient to meet the entire growth of coal demand well into the next century or beyond.

Unfortunately, however, this global superabundance of coal is no guarantee against the onset of major and long-lasting local shortages. This is partly because coal reserves are very unequally

G. ARMSTRONG

distributed around the world with the bulk remote from current markets, and partly because in an extractive industry there is a constant need, not always accepted in time, to locate economic reserves and to replace the mining capacity which is constantly being lost.

Movement of coal to the markets where imbalance occurs also presents problems. At present, out of a total world output of about 3×10^9 t of coal, only approximately 190×10^6 t are exported from one country to another to correct these local shortages. The bulk of these exports is from a small number of countries with those from the United States making up about 25 % of the whole. In view of its internal energy situation, it seems very unlikely that the United States will be able to increase, or even sustain, this level of exports in the future; while, despite adequate reserves, it is difficult to foresee the other exporting countries being able to increase their overall productive capacity to the degree needed to provide sufficient exports to significantly alter the basic imbalance that exists in many parts of the world.

In my opinion, it is therefore essential for each country blessed with economic indigenous coal reserves to develop these to the full, remembering the long lead time needed for detailed exploration and to generate new productive capacity. There will undoubtedly be marked changes in the supply pattern in the future, and locally there could be severe disruption due to incorrect anticipation or restrictions imposed by national monopoly interests.

Timely investment in world coal exploration and productive capacity becomes essential in view of the much more likely shortages which will undoubtedly affect the supply position of other energy fuels.

SELECTED REFERENCES (Armstrong)

Armstrong, G. 1972 Coal – the world's major fossil fuel. *Colliery Guard.* **220**, no. 11.
Averitt, P. 1967 Coal resources of the United States. *Bull. U.S. geol. Surv.* no. 1275.
Collins, H. E. 1969 Review of world coal production potential. *Colliery Guard.* **217**.
Schurr, H. S. 1963 Energy. *Scient. Amer.* p. 114.
Statistical Summary of the Mineral Industry 1967–1971 London: Institute of Geological Sciences.
Williamson, I. A. 1967 *Coal mining geology.* Oxford University Press.
World Coal Trade 1970 National Coal Association U.S.A.
World Power Conference 1968 Survey of energy resources.

Discussion

PROFESSOR J. M. CASSELS, F.R.S. (*Physics Department, Liverpool University*). If the market value of coal were to double overnight, by what factor would the National Coal Board's reserves be increased?

MR ARMSTRONG. Reserves would be increased by about 40 %.

Phil. Trans. R. Soc. Lond. A. **276**, 453–462 (1974)

Printed in Great Britain

453

Oil reserves and production

By Sir Eric Drake

The British Petroleum Company Ltd, London

The growth of world energy requirements over the last two decades has created an increasing demand for crude oil. Production has doubled in each of the decades to meet the demand and it is forecast that the demand for crude oil will continue to increase to 1985. The annual production rate has now reached the level of the average annual discoveries over the last 25 years, and the remaining proved recoverable reserves will probably decline continuously as production rates continue to grow. The declining reserves will be insufficient to support the forecast demand after about 1978 when the demand may be limited by the availability of crude oil. It is estimated that about half of the ultimate recoverable reserves have been found to date. The discovery of the remaining reserves will present difficult technical problems.

1. Introduction

Petroleum occurrences are widespread in the sedimentary basins of the world but petroleum accumulations that are large enough to justify production operations and hence classification as a reserve are less frequent. The minimum size of accumulation needed to justify production depends on the economics of the operations and reserves are therefore influenced by economic circumstances. The reserves, production and demand for oil discussed in this paper are based on the technical and economic circumstances foreseen in September 1973. The economic situation has been changed already by price increases introduced by the producer countries, one effect of which will be to make alternative fuels more competitive with oil and thus eventually tend to lower the demand for oil.

Reserves and production of the U.S.S.R., Eastern Europe and China are not considered in this paper, and unless otherwise stated the world reserves and production do not include the resources of those countries.

The proved recoverable reserves are the volume of oil remaining in the ground which geological and petroleum engineering information indicate with reasonable certainty can be recovered in the future from known reservoirs under existing economic and operating conditions.

The ultimate recoverable reserves are the total quantity of oil that will be recovered from sedimentary rocks of the world and are the sum of cumulative production to date, present proved recoverable reserves and estimated recoverable reserves still remaining to be discovered.

The reserves/production ratio is derived from the proved recoverable reserves at the end of a year and the annual production for that year. This ratio provides a useful relationship by which to study trends but it does not define the life of the reserves as it does not take account of the pattern of future production. A reserve/production ratio of 10 has become established by economics of the industry in the United States, but in the rest of the world where production developments are less rapid a minimum ratio of about 15, which implies a proportionately larger reserve base, is probably more realistic.

Coal and natural gas are converted to oil equivalents on the basis of their calorific value. Nuclear and hydro-electric power are converted on the basis of the heat value of the electricity generated.

2. Growth of world energy demands – excluding U.S.S.R., Eastern Europe and China

The total energy demand of the world increased at an annual average rate of 4.5 % between 1950 and 1970. The annual average growth rate of oil, natural gas and nuclear and hydro-electric power exceeded the growth rate of the total energy demand throughout the period and their share of the demand increased at the expense of solid fuels (figure 1; table 1).

TABLE 1. Growth of world energy demand, excluding U.S.S.R., Eastern Europe and China, 1950–1970

	1950	1960	1970	% change, annual average 1950–70
	10^{17} J (10^9 barrels oil equivalent)			
total energy	654 (10.7)	935 (15.3)	1610 (26.2)	+4.5
oil	220 (3.6)	410 (6.7)	852 (14.2)	+7.0
natural gas	79 (1.3)	165 (2.7)	330 (5.4)	+7.0
water power ⎱ nuclear power ⎰	12 (0.2)	24 (0.4)	37 (0.6)	+6.1
solid fuels	342 (5.6)	336 (5.5)	366 (6.0)	+0.2

The annual average growth rate of both oil and natural gas was 7 %, a rate of growth which doubles annual production every 10 years. The contribution of oil to the world energy demand requirement increased from 33 % of the total in 1950 to 54 % in 1970. Over the same period the contribution of solid fuels dropped from 53 % of the total in 1950 to 23 % in 1970.

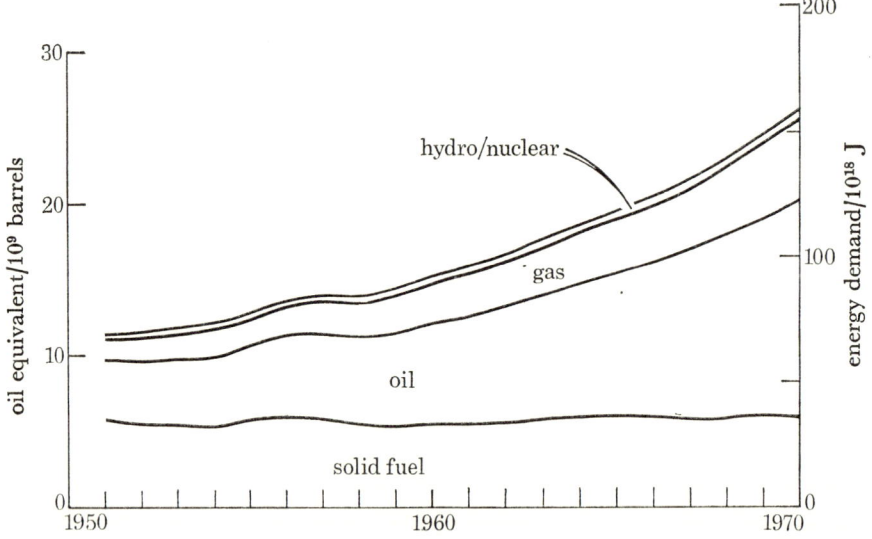

FIGURE 1. Growth of world energy demand, 1950–70, excluding U.S.S.R., Eastern Europe and China.

It is forecast that world economic growth for the next 12 years will continue at a slightly faster rate than in the last 10 years. The continuing population growth and demand for improved living standards will ensure sustained growth in demand for goods and services. Historically, energy demand has moved broadly in line with the growth in national output, though the energy requirement per unit of gross national product (g.n.p.) has tended to decline over time as economies reach an advanced industrial stage of development. For the world as a whole a continuing slight decline of the ratio of energy to g.n.p. is forecast. The range of annual energy consumption in 1970 from 320×10^9 J (the equivalent of 53 barrels of oil) per head in the United States to 60×10^9 J (10 barrels) for the whole of the world (figure 2) illustrates the potential for growth of demand that exists in the world.

In estimating the world energy requirements, forecasts of economic growth in each major area of the world have been made. While short-term fluctuations in supply and demand and Government intervention could change the energy demand, it is considered unlikely that the pattern of social and industrial useage of fuel will change significantly during the next 5 years. General awareness of a potential energy shortage may have a longer term effect on demand patterns which is difficult to quantify.

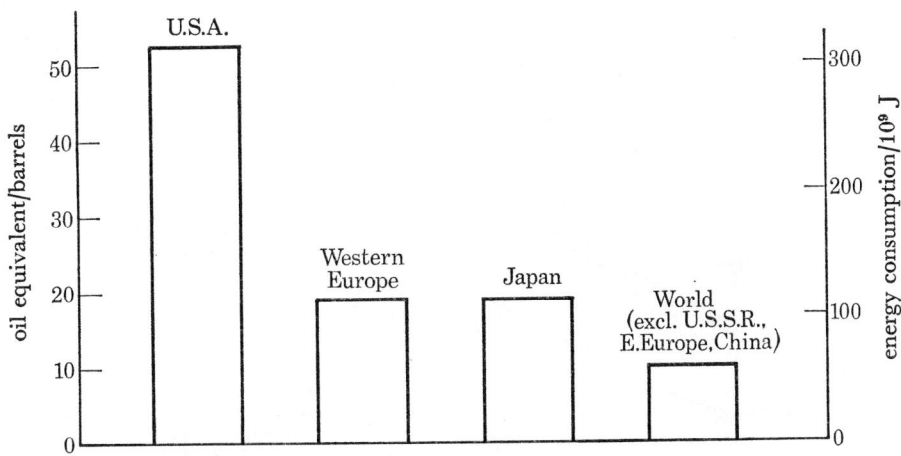

Figure 2. Primary energy consumption per head of population, 1970.

In forecasting the energy requirements to 1985 it has been assumed that economic growth will not be limited by energy supplies during the period under review, and the level of oil demand is forecast after taking into account the likely supply of other fuels (figure 3 and table 2). It is forecast that the world energy demand will grow at an annual average rate of 5.3 %, and after allowing for the likely growth of other fuels, the oil supply will be required to grow at an annual average rate of 5.5 % to satisfy the demand. If this demand is satisfied oil will therefore continue to provide an increasing share of the total energy demand; by 1985 this share is forecast to be 60 %. The increase is mainly at the expense of the solid fuels, which it is forecast will supply 16 % of the energy demand by that time.

TABLE 2. WORLD ENERGY DEMAND FORECAST 1972–1985, EXCLUDING U.S.S.R., EASTERN EUROPE AND CHINA

	1972	1977	1982	1985	% change, annual average 1972–85
	10^{17} J (10^9 barrels oil equivalent)				
total energy	1730 (28.2)	2240 (36.6)	2900 (47.3)	3380 (55.4)	+5.3
oil	980 (16.0)	1330 (21.8)	1750 (28.3)	2020 (33.0)	+5.5
natural gas	354 (5.8)	430 (7.0)	550 (9.0)	650 (10.6)	+4.8
water power	37 (0.6)	49 (0.8)	61 (1.0)	73 (1.2)	+5.0
nuclear power	6 (0.1)	31 (0.5)	67 (1.1)	104 (1.7)	+26.4
solid fuels	35 (5.7)	40 (6.5)	48 (7.9)	544 (8.9)	+3.4

To meet the potential demand it will be necessary for annual production rates to increase to 5.25×10^9 m^3 (33×10^9 barrels) by 1985, or approximately twice the annual production in 1972. However, the present production forecasts suggest that crude oil availability may limit demand by 1978 and thereafter oil scarcity is a real possibility. Also, if the present estimates of ultimate recoverable reserves are correct, it is unlikely that they will be able to support a production of 5.25×10^9 m^3 in 1985 and maintain a satisfactory ratio of reserves to annual production.

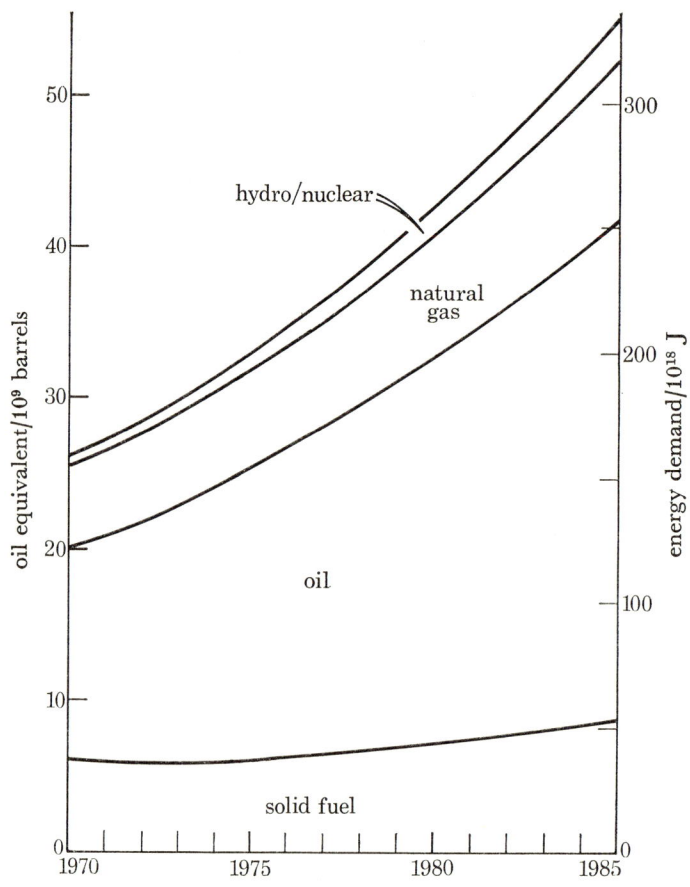

FIGURE 3. Forecast growth of world energy demand, 1973–85, excluding U.S.S.R., Eastern Europe and China.

3. PRODUCTION AND RESERVES

The world's resources of fossil fuel are finite in amount and have been generated in the course of geological time. They are now being consumed at rates that will lead to their exhaustion in a matter of hundreds of years, a negligible period on the geological time scale, and not therefore renewable within the time span of our industrial society.

In the case of petroleum, production has been increasing exponentially at a rate of 7 % per annum for several decades. This rate of growth doubles production in 10 years and also doubles the cumulative production during the same period. Production rates have now reached the level at which the cumulative production for a decade represents a substantial proportion of the world's estimated ultimate recoverable reserves and it is obvious that such growth rates can only be supported for a limited period. From the birth of the oil industry in 1859 to 1962 it is estimated that 18×10^9 m³ (114×10^9 barrels) of oil, or $7\frac{1}{2}\%$ of the present estimated ultimate recoverable reserves, were produced. By the end of 1972 a further $7\frac{1}{2}\%$ had been produced. Since 1859 oil production has expanded to meet the growing demand for energy but in the next phase of the production history the rate of increase of production must fall, eventually to zero, and thereafter production rates will slowly decline.

The nature of petroleum makes the estimation of the world's ultimate recoverable reserves difficult and various estimates have been published. The difficulties arise from the irregular distribution of reserves, which preclude any simple relation between reserves and sedimentary volumes or basin types, and the special geological circumstances that are necessary for an accumulation to be technically and economically producible. These circumstances occur within sedimentary basins from depths of about a hundred metres to more than 6000 m, which is the probable lower limit for significant oil accumulations. In the last 20 years a number of authorities (table 3) have estimated the world's ultimate recoverable reserves of oil to be between 160 and 400×10^9 m³ (1000 and 2480×10^9 barrels).

TABLE 3. ESTIMATES OF WORLD ULTIMATE RECOVERABLE RESERVES OF CRUDE OIL

year	source	reserves 10^9 m³	reserves 10^9 barrels
1949	Levorsen	238	1500
1949	Weeks	161	1010
1953	MacNaughton	159	1000
1956	Hubbert	199	1250
1958	Weeks	238	1500
1959	Weeks	318	2000
1965	Hendricks (U.S.G.S.)	394	2480
1967	Ryman (Esso)	332	2090
1968	Shell	286	1800
1968	Weeks	350	2200
1969	Hubbert	215–334	1350–2100
1970	Moody (Mobil)	286	1800
1971	Warman (B.P.)	191–318	1200–2000
1971	Weeks	364	2290
1972	Warman	302	1900
1972	Bauquis, Brasseur & Masseron (I.F.P.)	310	1950

It is noticeable that these estimates tend to increase from about 160×10^9 m³ in the early 1950s to Hendricks' estimate of 400×10^9 m³ in 1965, and have subsequently settled at about 300×10^9 m³ (1900×10^9 barrels). The estimates have been made during a period of worldwide and intensive exploration activity using increasingly sophisticated exploration techniques. Knowledge of the extent and prospects of most major sedimentary basins, both onshore and offshore, has been gained from geophysical surveys and well information, and it is reasonable to assume that the later estimates have been influenced by this information. In view of the widespread of present-day knowledge and the increased number of sedimentary basins in which exploration activities have been carried out, it is also reasonable to assume that continued expansion of the estimates of ultimate recoverable reserves is less likely to occur than in the past.

All the figures quoted by the authorities listed refer to the total world ultimate recoverable reserves. The reserves of the U.S.S.R., Eastern Europe and China may amount to some 71×10^9 m³ (450×10^9 barrels). On the basis of this estimate ultimate recoverable reserves of about 230×10^9 m³ (1450×10^9 barrels) is indicated for the rest of the world. Up to the end of 1972 it was estimated that 110×10^9 m³ (700×10^9 barrels) of oil, or nearly 50 % of the western world's ultimate recoverable reserves had been discovered. Of this total, 36×10^9 m³ (230×10^9 barrels) have been produced leaving present proved recoverable reserves of 75×10^9 m³ (470×10^9 barrels). This figure is lower than the reserves now quoted in industry journals, which are believed to include some over-optimistic figures. As it is likely that most of the more easily found reserves – those contained in the giant anticlines, the shallow structures easily detected by geophysical methods, the structures in shallow water – have already been discovered, the task of finding the remaining half of the ultimate recoverable reserves will obviously be more difficult.

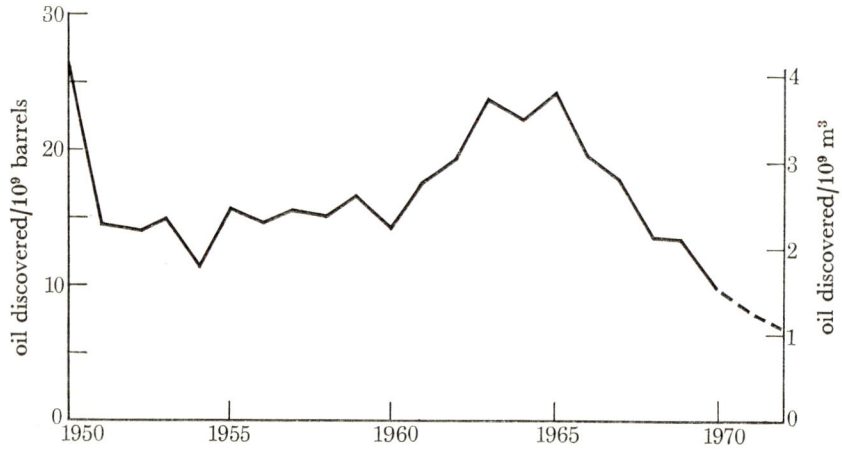

FIGURE 4. World oil discoveries, excluding U.S.S.R., Eastern Europe and China – 5-year running mean.

The problems that lie ahead can be seen from an examination of the historical discovery figures (figure 4). Annual discoveries vary widely from year to year and a five-year running average has been calculated in an attempt to indicate trends in discovery rates. Over the last 25 years, when much of the Middle East reserves were discovered, the average rate of discovery has been about 2.9×10^9 m³ (18×10^9 barrels) per annum. Although the figures for the last year or two are still subject to revision, they do extend the period in which discoveries have been below average and suggest that the previous discovery rate may not be maintained in the future.

The presently known reserves are extremely irregularly distributed, with 70 % of the total located in a comparatively small part of the Middle East, and the remaining 30 % divided fairly evenly between North America, Africa and the rest of the world (figure 5). The possible future discoveries may be more evenly distributed, with North America and the Middle East each containing about 20 % of the total and Africa about 10 %. The other countries of the world may contain 20 %, while the remaining 30 % may be in presently unforeseen areas and in deep-water prospects. As the technology to exploit the deep-water prospects is still under development, it is doubtful whether this portion of the reserves, if proved, can contribute to production within the period under review.

From 1959 to 1970 the annual crude oil production of the world, excluding the U.S.S.R., Eastern Europe and China, quadrupled from 0.57×10^9 to 2.25×10^9 m³ (3.6×10^9 to 14.2×10^9 barrels). Until 1968 annual additions to reserves generally exceeded production and the remaining proved reserves increased in this period from an estimated 48×10^9 to 78×10^9 m³ (300×10^9

to 490×10^9 barrels). Since that date production has exceeded the annual additions, which have been below the average rate of discovery over the last 25 years, and the proved reserves have now declined to 75×10^9 m^3 (470×10^9 barrels). Throughout this period the reserves/production ratio has declined steadily, due to the exponential increase in production rates against a more uniform average rate of discovery of new reserves. The forecast demand for 1974 is 2.86×10^9 m^3 (18×10^9 barrels), which is the 25-year average annual rate of discovery and it is therefore probable that future production will exceed additions to reserves and that the remaining proved recoverable reserves will decline continuously.

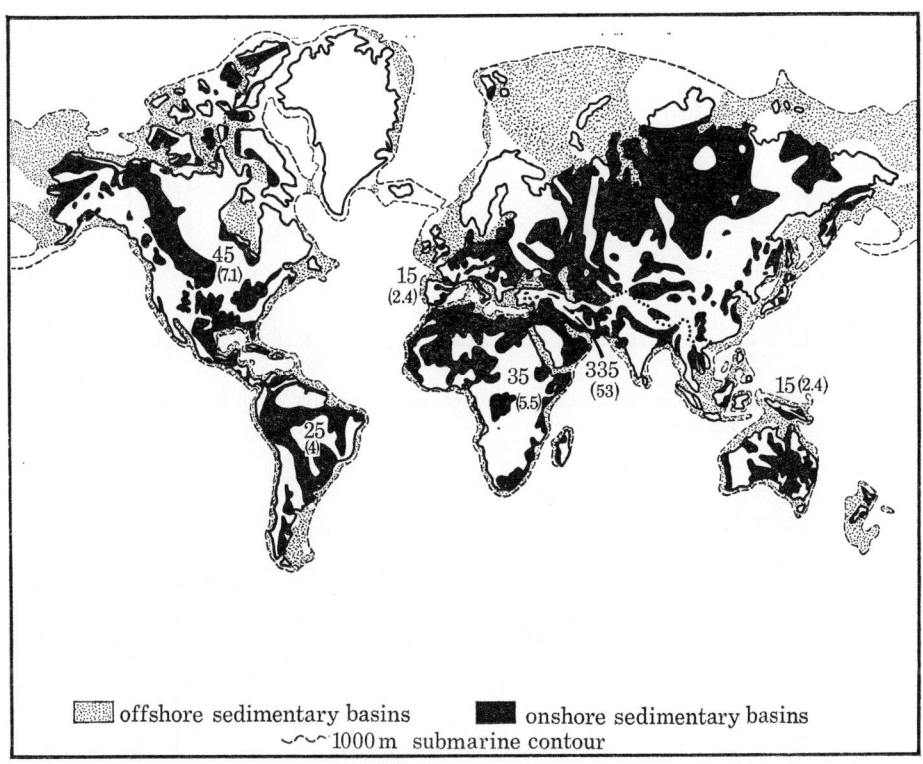

FIGURE 5. Distribution of remaining proved oil reserves – 10^9 barrels (10^9 m^3). Total remaining at end 1972 is estimated to be 470×10^9 barrels (75×10^9 m^3).

The proved reserves figures quoted above for 1950 to the present are substantially higher than the annually published figures for this period. The figures used here have been obtained by referring the reserves of major fields, that is fields with reserves of more than 80×10^6 m^3 (0.5×10^9 barrels) back to the year of their discovery. As the reserves contained in major fields constitute about 75 % of the proved reserves it is believed this procedure gives a realistic assessment of both the reserves and the reserves/production ratio.

The historical curves of annual production, remaining reserves and reserves/production ratio have been extended to 1985, showing both the potential demand and the production forecast (figure 6). The forecasts of production and demand over this period cover a range of values for both quantities. Mean values of supply and demand have been used to demonstrate the probability that demand, which is forecast to increase from 2.7×10^9 m^3 (17×10^9 barrels) in 1973 to 5.25×10^9 m^3 (33×10^9 barrels) in 1985, may be limited by supply after 1978.

In estimating the remaining reserves after 1973 it has been assumed that the increasingly

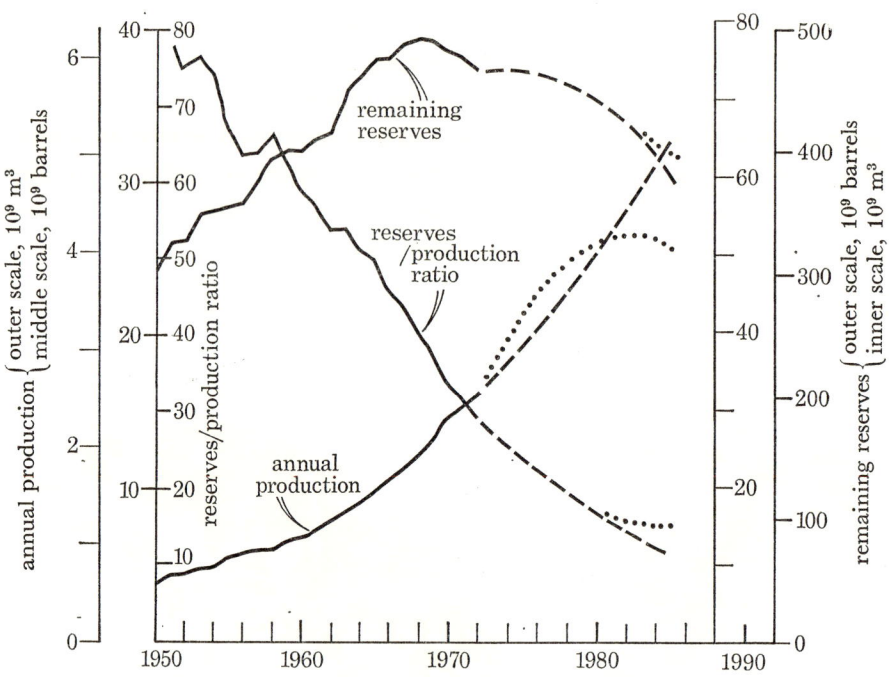

FIGURE 6. World production, reserves and reserves/production ratio, 1950–72 and forecasts for 1973–85, excluding U.S.S.R., Eastern Europe and China. — — —, forecast demand; . . . ., forecast availability.

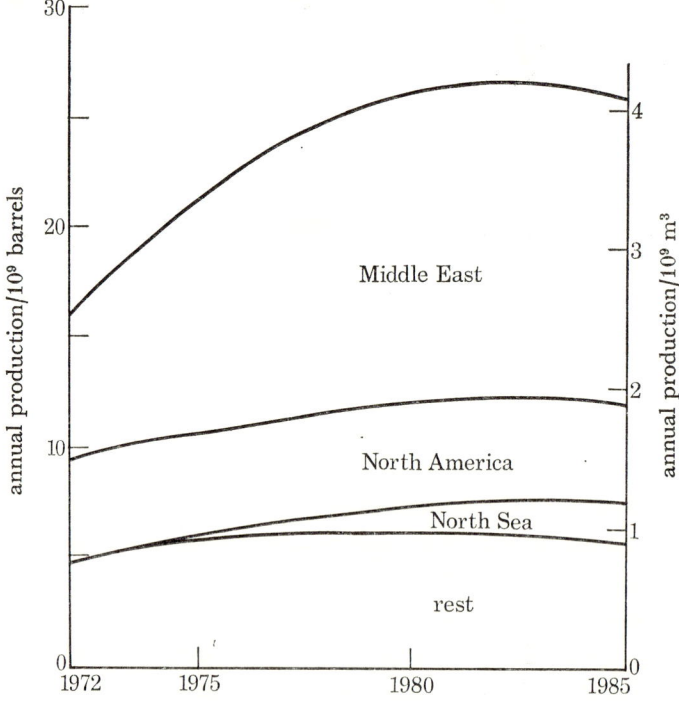

FIGURE 7. Forecast world crude oil production, 1973–85, excluding U.S.S.R., Eastern Europe and China.

sophisticated exploration techniques will maintain the 25-year average rate of discovery of 2.86×10^9 m³ throughout this period. In spite of this optimistic assumption, the reserves will decline to about 60×10^9 m³ (380×10^9 barrels) and the reserves/production ratio to 11 by 1985 if the demand forecasts are met. This ratio is considered to be too low and implies a reserve base that is too small to support the production required to meet the forecast demand. The limitation of annual production to 4.3×10^9 m³ (27×10^9 barrels) in 1982/83 followed by a slow decline of production will allow the reserves/production ratio to be maintained above 15, which is probably a more realistic level for worldwide operations.

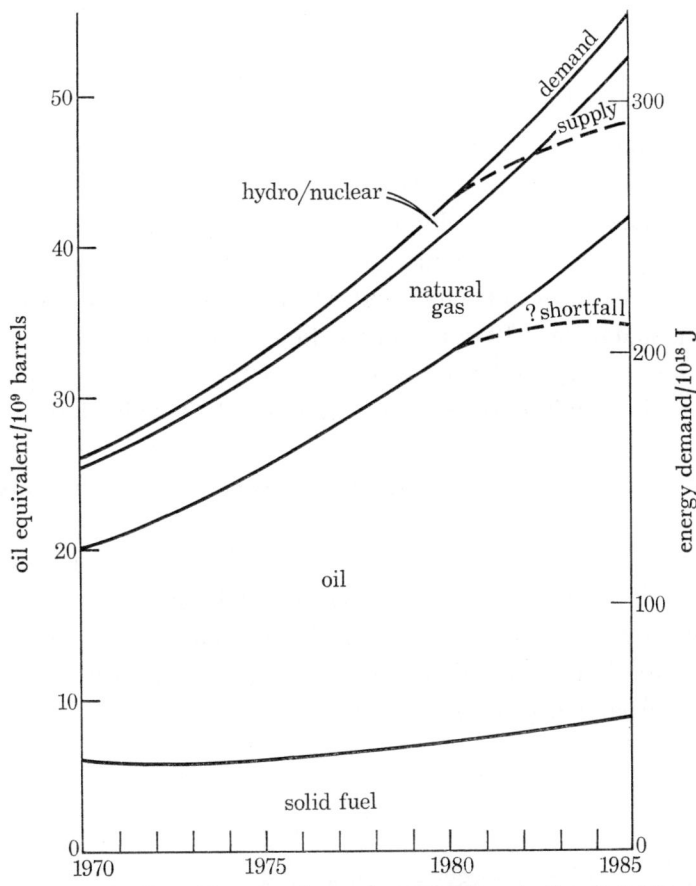

FIGURE 8. Forecast growth of world energy demand, 1973–85, excluding U.S.S.R., Eastern Europe and China, showing the effect of a possible oil shortage.

The production throughout this period will be heavily dependent on Middle East sources (figure 7). The contribution of this area to the total western world production is forecast to increase from 40 % in 1973 to 53 % in 1985. The North Sea production will amount to 7 % of the world supply in 1985.

The possible limitations of demand by a shortage of crude oil has serious implications for the world energy demand. In the case of the mean values of the production and demand forecasts the shortfall would represent 15 % of the total energy demand (figure 8) and the growth of energy available would be reduced from 5.3 to 2.4 %. It is apparent that alternative energy sources must be developed rapidly if the world's economic growth is not to be restricted by a shortage of primary energy.

Note added in proof (*March* 1974)

Since this paper was prepared there has been an unprecedented rise in the world oil prices. If the increased prices are maintained it is likely that the medium and long term demand for oil will be less than had been forecast but it is too soon to predict the extent of this reduction. It is also likely that higher energy prices will encourage the production of coal and that solid fuels will regain some of their lost share of the world's primary energy consumption, thereby reducing the demand for oil. The combined effect of increased prices and increased competition from alternative fuels will defer the time at which the demand for oil will be limited by its availability.

Discussion

Sir Peter Kent (*Natural Environment Research Council*)

In a written contribution, Sir Peter Kent wished to emphasize that the discovery situation which had been described in Sir Eric Drake's paper was not likely to be fundamentally changed by new exploration. There are strict physical limits to the possible range of oil occurrences – particularly downwards, for raised temperature at depth precludes survival of liquid hydro-carbons, a depth ranging down to 6000 m under favourable circumstances, but commonly less than 4600 m and as shallow as 910 m in some sedimentary areas.

Not only are seismic survey methods now greatly sophisticated, with an accuracy and pene-tration far exceeding the standards of only 15 years ago, but virtually all the world's sedimentary basins except for some remote Arctic regions have been subject to at least reconnaissance survey. A good deal is now known also about the deep oceans, and the information is essentially unfavourable.

There are undoubtedly large numbers of small oil-fields awaiting discovery, and probably other areas comparable to the North Sea (of high regional importance) but it now seems quite unlikely that further reserves of the order of the Middle East will be found. Additions of this size would be necessary if the energy demand escalating at the present rate until the end of the 1980s was to be met from hydrocarbon sources.

Phil. Trans. R. Soc. Lond. A. **276**, 463–483 (1974)

Printed in Great Britain

Natural gas

By C. P. Coppack

Shell International Gas Limited, London

This paper reviews the world's existing natural gas reserves and future expectations, together with natural gas consumption in 1972, by main geographic regions. The present and future supply and demand situation in each of the main natural gas markets, namely the U.S.A., the U.S.S.R., Japan and Western Europe, are discussed in the context of rising energy demand. The reasons are examined for the apparent paradox that, although now and for many years to come, the world has a theoretical abundance of natural gas reserves, in practice a number of important markets are already short or will soon be so. The paper will conclude by discussing the problems of meeting the developing future demand for natural gas with available indigenous production combined with supplementary supplies, including imports and manufactured substitutes.

1. Introduction

Compared with oil and coal, it is only recently that natural gas has become a commodity in international energy trade. Some important dates are:

1951 First major Canadian gas export by pipeline to the United States.
1964 First export of Groningen gas from Holland.
1964 First delivery of Algerian liquefied natural gas (l.n.g.) to the United Kingdom.
1968 First delivery of Soviet gas to Austria.
1969 First l.n.g. from Alaska delivered to Japan.
1970 First Iranian pipeline gas deliveries to the U.S.S.R.
1973 First Soviet gas deliveries to West Germany.

The reason for this comparative tardiness is quite simply that gas is more difficult to handle than oil or coal, and that the necessary technologies (larger diameter pipelines and l.n.g.), which are today taken for granted, have taken time to develop. One important consequence of this is that, although existing world gas reserves are plentiful, with good prospects of further discoveries, the major world gas markets (which are relatively few in number) are already, or are becoming, supply limited.

Looking ahead into the 1980s, although the flow of international trade in natural gas will most certainly intensify and multiply, it is unlikely, because of the delays which are inherent in the technological and political complexity of the many separate large projects which will be necessary, that such growth will be sufficient within this period to make any really effective change in the broad relation between world reserves and consumption which exists today.

A conclusion which can be drawn, therefore, is that there is a challenge which confronts the natural gas industry world-wide, together with the governments of the countries involved. It is to bring these abundant gas reserves economically, competitively and without undue delay to market, to the mutual benefit of both producer and consumer.

2. Existing world gas reserves

There are no generally accepted definitions of natural gas reserves and wide variations in both terms and estimating procedures may be met. For this reason what one country may classify as proven reserves cannot readily be compared with the proven reserves of another country, unless and until the definitions, quantification methods, energy values, composition, etc., are stated with some precision.

Further complications arise in quantifying exploitable reserves where gas is found dissolved in crude oil (solution gas) or in contact with underlying crude (gas-cap gas). Both are called associated natural gas and the volume of associated gas that can be recovered will depend mainly on the rate and the extent of oil production.

Because of these and other complications any world tabulation of gas reserves by country will almost certainly contain a wide variety of inconsistent data. However, these inconsistencies do tend to balance each other out to some extent and, in any case, are of less importance in the world picture than the relative broad regional disposition of world gas reserves.

With the foregoing reservations in mind, figure 1 indicates that known world gas reserves amount to some 53×10^{12} m³, of which about one-third is located in the U.S.S.R., with one-sixth each in the Middle East and North America.

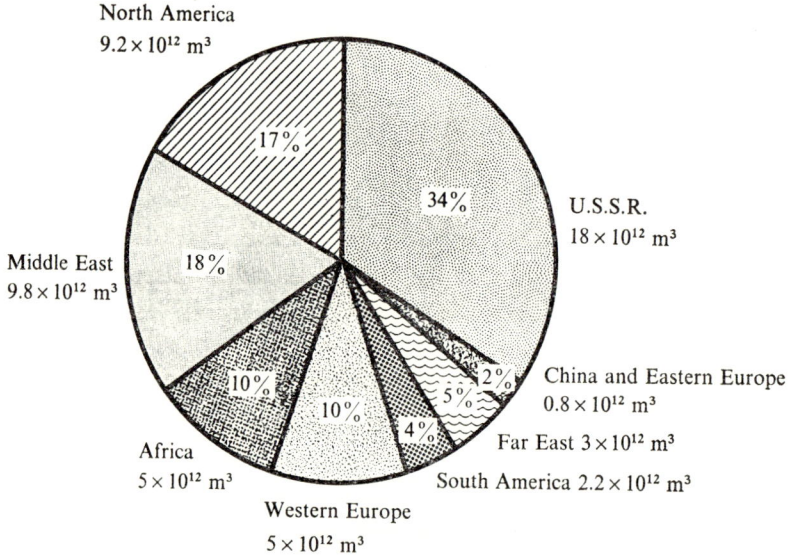

Figure 1. World gas reserves in 1972 – 53×10^{12} m³ (equivalent to about 330×10^9 barrels of oil or 2×10^{21} J) (source of data: *Oil and gas journal*, December 1972).

On a thermal basis world gas reserves are equivalent to somewhere between one-half and one-third of world reserves of conventional crude oil (i.e. excluding tar sands and shale oil). Oil currently supplies about 45 % of world energy demand and is also the main source of primary energy in virtually every country. Natural gas supplies some 17 % of world energy demand (at first sight, a reasonable proportion in view of the relation between oil and gas reserves), but unlike oil its use, as will be shown, is concentrated in a relatively few countries. This limited geographic spread of gas consumption is because many gas reserves are distant from markets and because gas is more difficult and more costly to transport than oil. The absolute volume of world gas reserves is in fact overshadowed in present-day effect and importance by the relative proximity of such reserves to the main consuming markets. Thus, gas reserves in North America and Europe have been easier to develop and bring to market and hence have had a greater effect so far upon the pattern of world energy supplies than reserves in the Middle East.

In the future, as international trading in gas evolves from its present beginning, distant gas reserves will become economically linked to the world's major markets and a relation between world reserves and world demand will become more meaningful.

3. EXPECTATIONS FOR FUTURE DISCOVERIES

With the notable exceptions of North America, the U.S.S.R. and Europe, natural gas has until recent years been much less attractive as an exploration objective than oil. With the emergence of international trading in gas and the increasing awareness of the need to maximize all energy forms, the position has changed and new gas discoveries may be economically attractive in many other areas.

Estimates of the future rate of discovery of gas reserves are fraught with precisely the same uncertainties as the assessment of future discoveries of oil or, indeed, any other minerals. Some geologists have predicted that ultimate recoverable gas reserves are likely to be of similar magnitude (on a thermal basis) to ultimate recoverable oil reserves, while others have been less optimistic. The indications given in figure 2 of the possible level of gas reserves that remain to be discovered should be regarded as being no less or more accurate than any other responsible prediction.

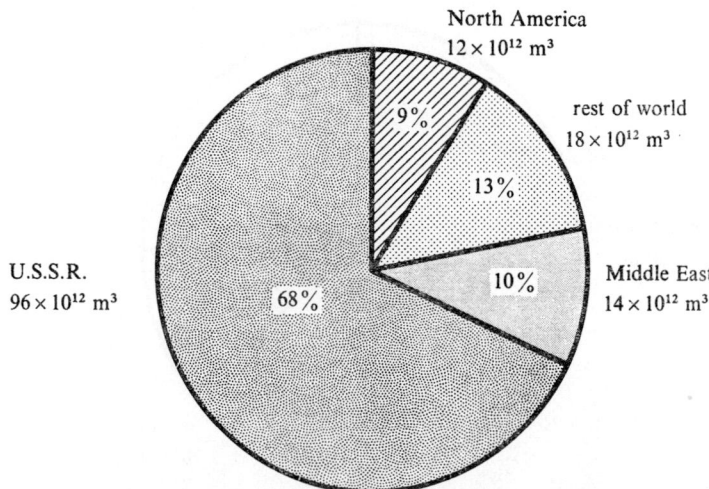

FIGURE 2. Expectations from new discoveries in addition to proven reserves – 140×10^{12} m³ (equivalent to about 860×10^9 barrels of oil or 5.25×10^{21} J).

Quite apart from the actual quantities of gas that remain to be discovered, which can be determined only after an exhaustive search over many years, there is the further imponderable of the extent to which economic incentives will exist to search for and develop all gas-bearing regions, particularly those situated in difficult locations or remote from the main consuming markets. Some sizeable gas deposits may be discovered but never in fact be developed because of economic considerations, technical complexity or because over time the potential market outlets for the gas may disappear as demand turns to other newer energy forms such as nuclear, hydrogen, etc. Forecasts of future gas discoveries should in addition be hedged with the same reservations as those already discussed in regard to existing reserves, that is to say that the contribution which new gas discoveries are likely to make to future energy demand will remain critically dependent on their location, the rate of development of the technology of gas transportation and (perhaps most important of all) the extent to which political, economic and commercial factors can be brought harmoniously together so that the necessary investments are made and facilities constructed.

So far this discussion has stressed the complexity of relating world reserves, both existing and future expectations, to demand. It is perhaps appropriate now to attempt to clarify the picture by examining where the demand for gas exists today and how it might be expected to develop in the years ahead.

4. World gas consumption

Historical data of world gas consumption are better documented and more precise than reserve data. Although the statistics for 1972 are still as yet incomplete, figure 3 is nevertheless thought to be a fairly accurate portrayal of world gas consumption by main region for that year. It shows that North America, essentially the U.S.A., is the world's largest gas market and that the U.S.A., the U.S.S.R. and Europe together account for nearly 90 % of the world's total gas consumption. Excluding China and Eastern Europe, the balance of 6 % for the 'rest of world'

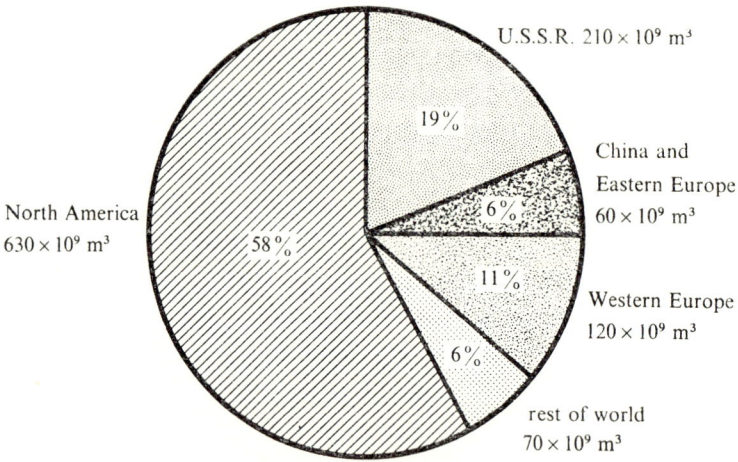

Figure 3. Estimated natural-gas consumption in 1972 – 1090 × 10⁹ m³
(equivalent to about 7 × 10⁹ barrels of oil or 43 × 10¹⁸ J).

represents a large number of small markets, e.g. Argentina, Australia, Bangladesh, New Zealand, Nigeria, Pakistan, Trinidad, Venezuela, etc., where natural gas is important in the local context but not in world-wide terms. Somethat different considerations, however, apply to Japan, one of the world's largest energy consumers. This is a relatively new market for gas which will become of increasing importance in the years ahead; the prospects for gas in Japan are discussed separately later in this paper.

Comparison of figures 1 and 3 shows some substantial regional variations between indigenous reserves on the one hand and consumption on the other. For example, whereas the U.S.A. is by far the largest gas market in the world, its share of more or less proven world reserves is at best only about 17 %, ranking third after the U.S.S.R. and the Middle East.

For the world as a whole the reserves : consumption ratio for natural gas (derived from figures 1 and 3) is just under 50:1, from which it might reasonably be concluded that there is an ample reserve base to satisfy expected market demands for many years ahead. However, as already mentioned, the practical, physical, political and commercial problems of developing and bringing to market a substantial proportion of these gas reserves are such that, for gas, global reserves : consumption or production ratios are not suitable yardsticks for measuring the true extent of the exploitable resource base in relation to present market demands. It is all a matter of timing; in due course a large part of these reserves will almost certainly flow to the world's markets, but today the necessary transportation systems are still being developed or (perhaps in certain cases) have yet to be invented. It would seem unlikely, however, that within the period under discussion (the 1980s) the development of such systems will be sufficient to change very substantially the broad picture as it exists today of ample world reserves coexistent with supply-limited markets. We thus have the apparent paradox that at this time, and for many years to come, the world has an arithmetic abundance of natural gas resources, while in practice a number of important markets are already experiencing or will soon experience supply shortages. Let us now look at the reasons for this in greater detail by examining the situation separately for each of the main markets of interest.

5. The U.S.A.

It is worth while considering in some detail the American natural gas scene, as many of the lessons which are being learnt in this market can in due course be applied elsewhere in the world where similar problems of demand outstripping supply are beginning to emerge.

For many decades the United States has enjoyed abundant supplies of indigenous energy, that is to say of coal, oil and natural gas. These resources have contributed significantly to the country's economic growth, national security and quality of life. In more recent years, however, development of new indigenous reserves of fuel have not kept pace with the growth in demand. A continued and increasingly severe deterioration of the energy supply position is predicted in the immediate years ahead by virtually all authoritative bodies both governmental and private.

In the case of natural gas, since 1966 annual production from the lower 48 states has been greater than new additions to reserves, with the consequence that the reserves : annual production ratio has declined from about 17:1 in 1966 to less than 11:1 in 1972. In other words, demand has been increasing appreciably faster than new additions to reserves. This statement still holds good even with the addition of Alaskan reserves, though these reserves are not yet in fact available to the market. The energy supply situation is exacerbated by the fact that oil demand, like gas, is also outstripping indigenous supply. For example, the U.S. Department of the Interior predicts overall U.S. demand for oil during 1973 at 2.8×10^6 m^3 (17.5×10^6 barrels)/day of which over a third, or more than 0.95×10^6 m^3 (6×10^6 barrels)/day will have to be imported. In the years ahead not only will demand rise but so will the percentage and hence absolute quantities of imported oil required.

What then has led to this situation? Figure 4 shows the historical development of the U.S. primary energy consumption from 1950 to today, together with a forecast for the next few years ahead.

The average annual growth rate in consumption of gas between 1950 and 1972 was 6.2%, compared with 5.3% for oil; coal use actually declined between 1950 and 1960 and has since staged a recovery to much the same level it enjoyed in 1950. This high rate of growth of gas is a reflection on the one hand of its inherent clean and versatile qualities as a fuel, and on the other of the low selling prices imposed by the Federal Power Commission (F.P.C.), which is the governmental regulatory body for all inter-state natural gas activities. The average price of gas at wellhead for inter-state movements is still only about U.S.¢19/10^9 J (¢20/10^6 Btu). Within this average the wellhead price for new gas dedicated to the inter-state market was still as low as ¢25/10^9 J (¢26/10^6 Btu) in 1972, and only very recently and under certain special circumstances have prices for gas moving in inter-state commerce been free to rise to a level equivalent to those of other fuels. Gas sold at ¢25/10^9 J at the wellhead and delivered to New York is sold there at less than one-half the cost of oil. It is no wonder then that gas, quite apart from its many desirable physical qualities, has been and still is the preferred fuel.

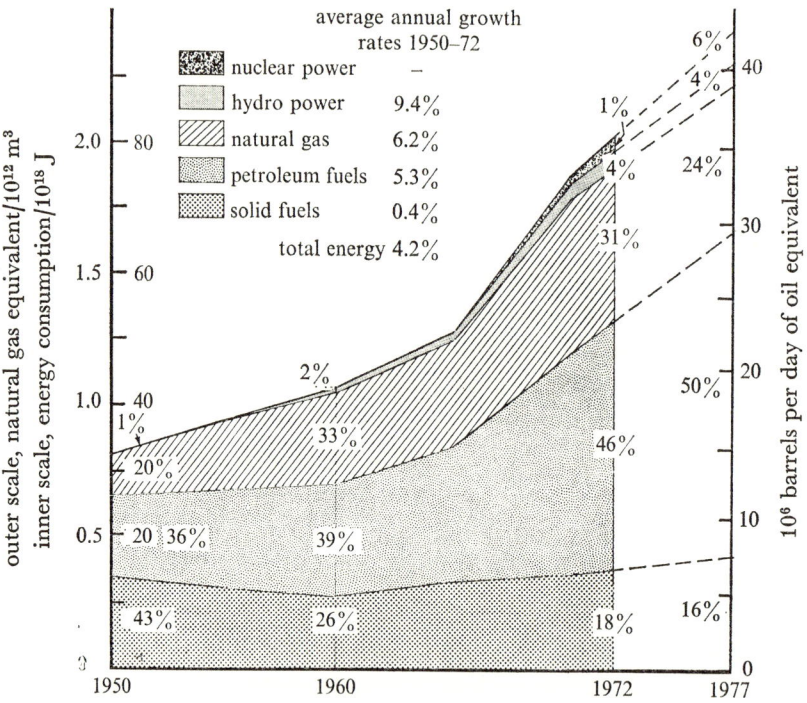

FIGURE 4. U.S.A., inland energy consumption and forecast, 1950–77
(excluding ships' bunkers and international aviation).

The consequence of this regulated low pricing policy has been that large quantities of gas have flowed into the intra-state market which is not subject to the same degree of stringent regulation, and into non-premium uses of gas such as for under-boiler fuel. More importantly, the policy has offered little incentive for the industry to explore for and develop new gas reserves for the predominant inter-state market.

A number of reports have been published by American study groups which conclude that a great deal more gas remains to be discovered in and around the United States. It is generally agreed, however, that much of this new gas is likely to be located on land at depths greater than 4500 m, in offshore structures, or in Alaska, and thus will be expensive (in the context of today's U.S. price structure) to find. Considerable exploration work is necessary before the extent of these new reserves can be determined with any degree of accuracy, but a reasonable assumption is that such new discoveries (excluding Canada) could reach as much as 9×10^{12} m³ or very roughly the equivalent of the total presently proven reserves of 9.2×10^{12} m³ in North America, i.e. including Canada. What is quite certain is that it will be costly and time-consuming to explore for and prove such reserves. Even if the appropriate financial incentives are forthcoming and new areas are made available for exploration fairly quickly, such new gas can even when found have little impact on overall supply availability for a number of years to come.

These supply constraints are beginning to be recognized by the U.S. Government, and President Nixon's Energy Message to Congress of April 1973 included proposals that indigenous gas from new wells, gas newly dedicated to the American inter-state market, and the continuing production of natural gas when contracts have expired should no longer be subject to price regulation at the wellhead. He also proposed the tripling of federal lease sales to stimulate new exploration effort. Moreover, in his State of the Union message in September 1973, the President urged Congress to take 'decisive action this year' on these matters. He proposed that 'we begin a gradual move to free market prices for natural gas by allowing the price of new supplies of domestic natural gas to be determined by the competitive forces of the market-place'.

How quickly and even to what extent these proposals may be approved by Congress remains to be seen, but in the present political climate it may be optimistic to expect an early resolution of these matters. Meantime because of supply difficulties an increasing number of gas distributors are having to curtail their sales in a variety of ways, such as by non-renewal of low-price contracts, refusal to accept new customers, etc. These actions, of course, impose a further burden upon the demand for other fuels and in particular for oil.

Apart from the discovery of new reserves, what are the other options open to the American gas industry? Basically these fall under two categories – imports of l.n.g. and the manufacture of synthetic natural gas (s.n.g.) from oil and coal. However, even under the most optimistic assumptions these supplementary sources of supply can in no way provide a sufficient supply base for gas to maintain the one-third share of total energy demand that it enjoyed in the late 1960s.

Taking s.n.g. from coal first, the problem here is not the availability of the resource base – the U.S.A. has vast deposits of coal much of which can be easily mined – but the technological complexity of converting coal to a high-energy gas compatible with natural gas. A substantial effort is being brought to bear upon this problem and a number of processes have reached the pilot-plant stage in the U.S.A. and a small-scale plant was recently commissioned in the U.K., but in spite of these encouraging developments it seems unlikely that large-scale commercial-size plants (say plants each capable of producing upwards of 2×10^9 m³/year of s.n.g.) can be developed and brought into operation much before the late 1970s/1980. One reasonably realistic assessment of the contribution which s.n.g. from coal could make to gas availability in the U.S.A. is that 20–25 plants costing together some 5×10^9 to 6×10^9 could be on-stream by the mid-1980s, producing a total of around 60×10^9 m³/year of s.n.g. (say 6×10^9 ft³/day). The estimated cost of s.n.g. produced from coal must inevitably still be very tentative at this early stage of technological development. Current estimates based on coal prices of about $8/tonne indicate ex-plant prices for s.n.g. at a high (92 %) load factor of at least U.S. ¢133/10^9 J (¢140/10^6 Btu) but commercial reality, when processes have been fully tested and built, may result in appreciably higher prices. To these prices will of course have to be added the cost of transporting the gas to market. Another factor to be taken into account in trying to predict the size of this source of supply is that since much of the coal required will have to be produced by open-cast mining, there may well be serious resistance on ecological grounds to massive developments on the scale mentioned above. For example, the production of 60×10^9 m^3/year of s.n.g. from coal could require as much as 170×10^6 tonnes of coal per year. If this were all or predominantly open-cast coal, and assuming 2.5 m thick coal seams, 70 km^2 of land would be need to excavated (and presumably rehabilitated) each year. The contribution that s.n.g. from coal can make to the U.S. energy demand may well therefore be constrained by technological and ecological problems, and perhaps also by price, in spite of the abundance of the raw material potentially available and in spite of the environmental advantage of producing a clean-burning fuel from coal containing varying proportions of sulphur. It is, nevertheless, likely to be an important contribution, the more so because it depends upon a secure source of supply.

Unlike s.n.g. from coal, commercial processes already exist for the manufacture of s.n.g. from oil fractions and such plants can now be built with confidence and brought on-stream in about two years from the beginning of construction. The constraint here is not of technological difficulty but rather that the feedstocks, naphtha-type fractions and gas liquids in particular are becoming increasingly expensive in line with all other oil products. This situation is aggravated by the fact that the ideal oil fractions for gas making are precisely the same fractions that are required as feedstock for the expanding petrochemical industry. Since the capital investment per joule of gas produced is lower for oil-based s.n.g. plants than it is for l.n.g. imports and for coal gasification, naphtha gasification may temporarily be employed to fill short-term needs, but as naphtha prices increase so the balance swings in favour of s.n.g. from coal and l.n.g. A substantial proportion of the naphtha required in the U.S.A. for gasification would have to be imported. Over 35 oil gasification proposals have been announced so far, of which one has been completed and a further seven are under construction. Whether the remaining proposals will materialize, and indeed many others yet to be announced, will depend upon the promoters' ability to secure a long-term supply of feedstock and, in appropriate cases, the necessary approvals from the F.P.C. As an illustration of the quantities that could be involved, over 95000 m^3/day (600000 barrels/day) of naphtha-type fractions would be required to produce some 25×10^9 m^3/year of s.n.g. (say, 2.5×10^9 ft^3/day).

Because the cost of making s.n.g. from oil is very sensitive to the price paid for the feedstock, the ex-plant cost of s.n.g. from oil can vary considerably, depending upon what assumptions are taken regarding future oil prices. Based on present price trends for oil, ex-plant gas prices could well rise to (U.S. ¢166/10^9 J (¢175/10^6 Btu) or more by the late 1970s.

It has been shown that the contribution that newly found natural gas and s.n.g. (from oil and from coal) can make to the American demand for gas in the years ahead may be limited by the long lead times required for exploration and subsequent development if the exploration is successful, by ecological problems, by inadequate and/or unproven technology, and by the cost of and limited availability of suitable feedstocks. Equally, imported l.n.g. is not the saviour. It too is constrained in its potential contribution to total gas supply.

Although a number of l.n.g. proposals have been announced, and are at varying stages of planning, only one major base load l.n.g. import scheme has (at the time of writing this paper) received all the necessary governmental consents and started construction. This is the first of El Paso's schemes from Algeria involving the import of some 10×10^9 m³/year and starting in 1976. Other proposals have yet to receive F.P.C. approval and are still in the negotiating/planning stage. Nevertheless, in spite of this apparently slow start, it is probable that the U.S. will in due course be importing l.n.g. not only from Algeria but also from Nigeria, the U.S.S.R., the Middle East and perhaps Venezuela, Indonesia, South East Asia and elsewhere. The starting times for such projects are likely to fall between the late 1970s and the late 1980s. Depending upon what assumptions are taken regarding the attitude and steps the U.S. government will take to combat the developing energy shortage, the degree of cooperation that will exist in the future between the governments concerned, the economic viability of the projects for the commercial interests involved, and the availability of the necessary hardware (both plants and ships) and finance, it is possible to postulate a spread of likely l.n.g. import programmes for the U.S. ranging between 50×10^9 and 120×10^9 m³/year by the late 1980s. Certainly the resource base in terms of gas reserves surplus to local needs exists in most of the supplying countries mentioned to support an even higher level of U.S. imports; however, experience so far indicates that the time required to put together complex international projects of this magnitude is such that these higher levels seem unlikely. Quite apart from the time needed for the design and construction of the necessary facilities, commercial and financial negotiations are time-consuming and the obtaining of the necessary governmental consents (at both ends of the supply chain) is inevitably a major limitation. As for prices, depending on the distance from the supply source to the import point and the volumes involved, the delivered price of l.n.g. imports into the U.S. can vary over a wide range. But even for the long-distance supply sources l.n.g. landed cost is still likely to be competitive with that of producing s.n.g. from oil and coal, and indeed with pipeline gas from Alaska. The techniques of liquefying natural gas and of ocean transport of l.n.g. are now proven on a commercial scale, and based on recent operational experience and allowing for capital and operating cost inflation in the years ahead a likely range of prices for l.n.g. delivered at U.S. seaboard in 1980 is between U.S. ¢95 and 142/10^9 J (¢100 and ¢150/10^6 Btu).

Perhaps the two main points to make in regard to the U.S. gas market are, first, that it has the capacity in the future to absorb the maximum levels one can reasonably envisage of all forms of supplementary gas including not only l.n.g. and s.n.g. but also any new indigenous gas that may be discovered on land or offshore, and yet still have an overall shortage in relation to the potential demand. This point is illustrated in figure 5, where a forecast is made for U.S. gas supply ahead to 1990.

Secondly, because supplies will be insufficient and new gas will be expensive, the marketing pattern must and will change. Low-priced large-volume steam-raising-type applications will be phased out and supplies will tend to be directed at those market sectors which can afford to pay higher prices and where the inherent qualities of natural gas can be used to the maximum advantage. This in return will create load-balancing problems for the pipeline systems of the gas distributors and impose a greater burden on other energy forms to fill the gaps that gas has historically supplied.

While it may be considered that these observations tend to paint a pessimistic outlook for gas in the U.S. market, it should be noted that even with these supply constraints the U.S. will remain for many years to come the world's largest gas market. Equally, it should be kept in mind that these problems offer a major challenge to meet which the entrepreneurial skills and capacity for technological effort and innovation of the industry are certainly being mobilized with support from the government; a struggle is under way, the outcome of which may differ, markedly from that predicted here on the basis of facts available today.

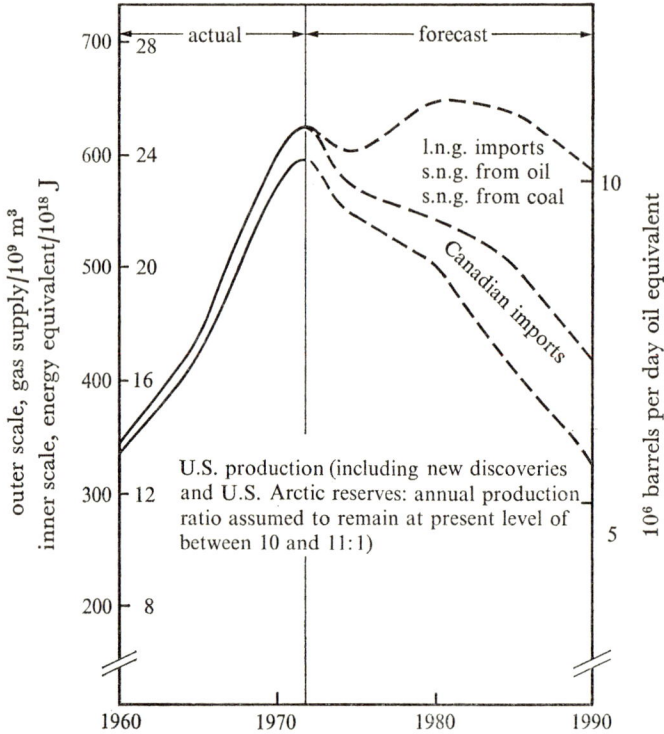

FIGURE 5. U.S. gas supplies – actual and forecast availability.

6. THE U.S.S.R.

In spite of the somewhat limited published data available, it is widely recognized that the U.S.S.R. has a natural gas resource base greatly in excess of its very large existing and future internal needs. Figure 6 shows how the contribution of natural gas to the U.S.S.R.'s energy consumption has grown from about 4 % in 1950 to a healthy 24 % by 1972 and is expected to continue growing both in absolute terms and percentage contribution through the 1970s.

Figure 1 indicated that current proven gas reserves in the U.S.S.R. are around 18×10^{12} m³ or about one-third of the best estimate of total world gas reserves. However, prospective U.S.S.R. reserves could be considerably greater. For example, a paper presented by a Soviet delegate to the 12th World Gas Conference at Nice in June 1973 (as also in other papers to various international bodies) indicated that prospective reserves, located at depths of less than 4500 m, could amount to 86×10^{12} m³, of which one-half are in West Siberia including the Tyumen area (see table 1 for details).

How much of these prospective reserves can be proved, developed and brought to market must be a matter for conjecture. The problems of drilling and constructing the necessary pipe-line systems in areas such as Tyumen, which are not only remote from the market and from suitable export points, but which also have climatic and topographical conditions which are among the most severe to be found in the world, are formidable by any standards.

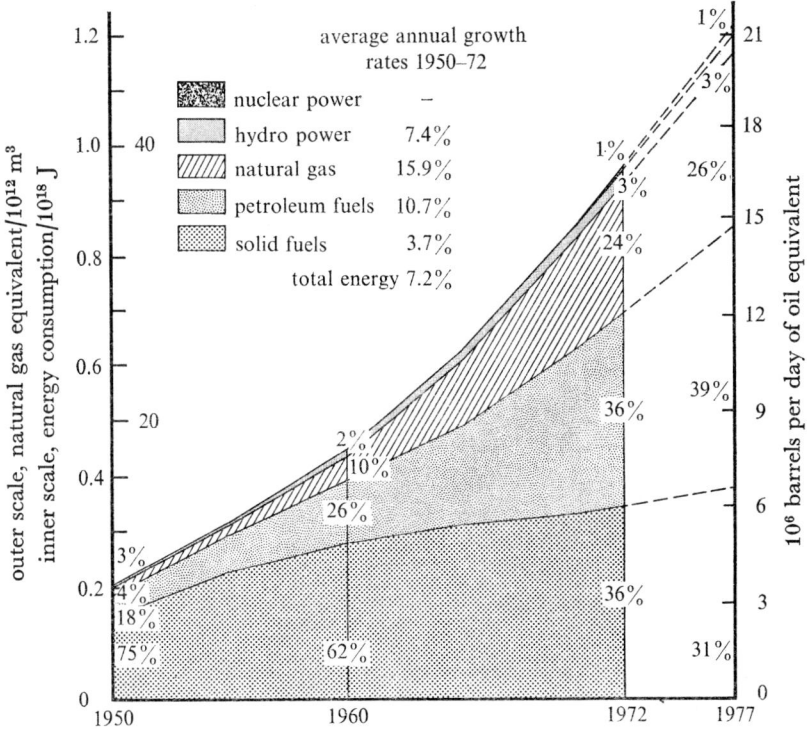

FIGURE 6. U.S.S.R. – energy consumption and forecast, 1950–77.

TABLE 1. GEOGRAPHIC DISTRIBUTION OF U.S.S.R.'s PROSPECTIVE GAS RESERVES

location	less than 4500 m		4500–7000 m	
	10^{12} m³	%	10^{12} m³	%
European U.S.S.R. (including Komi and Oldenburg)	15.0	17.4	5.6	56
west Siberia (including Tyumen)	42.3	49.2	0.2	2
east Siberia (including Sakhalin and Yakut)	15.7	18.3	–	–
middle Asia and Kasakhstan (including Turkmenian)	13.0	15.1	4.2	42
total U.S.S.R.	86.0	100	10.0	100

However, during the last few years the U.S.S.R. has undertaken long-term export commitments to supply natural gas by overland pipeline to Austria, Finland, France, Italy and West Germany, as well as to various East European countries, and negotiations are reported to be well advanced for possible supplies to Sweden. Gas to meet these commitments will come largely from existing fields in reasonably accessible locations that are either close to or already linked with the existing pipeline network. Even so this network does not have sufficient capacity to handle all these commitments and most of the U.S.S.R.'s gas deals so far have involved the reciprocal supply of high-pressure, large-diameter steel line-pipe.

More recently serious interest has been shown by various American gas companies, supported by the U.S. government, in the possible export of Soviet l.n.g. to both the U.S. east and west coasts. These export possibilities would require the development of west Siberian reserves in the case of supplies to the U.S. east coast and east Siberian reserves for the U.S. west coast. In addition to the massive capital investment required for liquefaction plants and l.n.g. ships, such exports would necessitate a large appraisal and development drilling programme, and the construction of new pipeline systems over long distances to suitable year-round ice-free export ports. The realization of these plans must depend upon the continuing political desire of both the American and Soviet governments to establish closer trading links between the two countries; if this is maintained, it can be assumed that the technological strength of the American and Soviet gas industries will surmount the practical problems. Even under the most favourable conditions, however, it will obviously be some years before such exports can actually start.

Whatever discount factor may be applied to the announced Soviet expectations of future gas reserves, the U.S.S.R. does and will have the necessary resource base to fulfil its present export commitments and to undertake substantially greater ones. Soviet gas is and will remain a most important factor in the world scene.

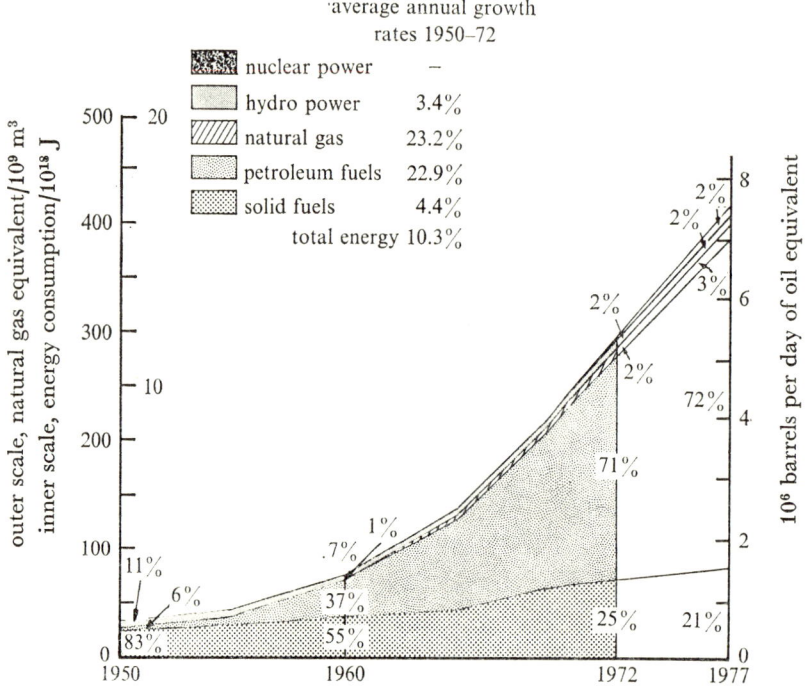

FIGURE 7. Japan – inland energy (excluding ships' bunkers and international aviation), consumption and forecast 1950–77.

7. JAPAN

The growth in energy consumption in Japan over the last 15 years or so has been quite remarkable by any standards, with total energy consumption rising from less than 2×10^{18} J (50×10^9 m^3 of natural gas equivalent) in the mid-1950s to about 12×10^{18} J (300×10^9 m^3) in 1972 (see figure 7).

This represents an average annual growth rate of over 10 % in total energy consumption and is the highest growth achieved by any industrialized country in the world. In the same period there have also been some radical changes in the energy supply pattern. For example, whereas in the mid-1950s coal supplied around 70 % of total energy consumption and oil little more than 20 %, by 1972 the percentage contribution of these two main primary forms of energy had been almost exactly reversed.

Japan has only very small indigenous reserves of natural gas and as can be seen from figure 7 natural gas's contribution to this growth in energy consumption was negligible. Natural gas represented barely 1% of primary energy consumption in the 1960s and only about 2% in 1972. It is possible, of course, that further reserves of natural gas may be discovered in the new exploration area now being explored offshore Japan, but in the absence of any such commercial discoveries, Japan must rely upon imports or s.n.g. if the gas market is to expand. Indeed the practice of importing natural gas in liquefied form is already well established. Since 1969 Japan has been importing 1.5×10^9 m³/year of gas in the form of l.n.g. from Alaska (the Phillips/Marathon project) and this was followed in December 1972 by the start of imports from Brunei (the Shell/Mitsubishi/Brunei Government project) in quantities which will rise to about 7×10^9 m³/year by the mid-1970s. These two l.n.g. schemes, together with the existing local production, will enable natural gas to supply about 3% of Japan's growing total energy demand by the late 1970s.

What then are the further prospects for natural gas utilization in Japan? First, the energy requirements of Japan are so large that virtually any quantities of gas (or oil) that may be discovered on land and/or offshore Japan would be readily absorbed by the market, bearing in mind the time it would take to develop such new reserves. As Japan relies on imported energy to meet well over 80% of its total requirements any indigenous supplies would help to reduce the country's dependence on foreign sources of supply in a world where oil availability is expected to become increasingly tighter. Secondly, Japan has a pressing need for clean energy to alleviate the very severe problems it is now experiencing as a consequence of its rapid industrial expansion and the concomitant increase in the standard of living and purchasing power of its population. Thirdly, based on the small indigenous supplies of natural gas and the existing imports from Alaska and Brunei, some of the gas distribution companies are converting or planning to convert their existing low-energy high-cost manufactured gas systems to high-energy natural gas. Over time this will improve their distribution economics and enable these companies to penetrate the space heating and other large-volume marketing outlets. Coupled with this is the increasing use of natural gas for industrial applications and for power generation either for reasons of fuel efficiency, for economic considerations or to reduce air pollution.

The above and other related factors all tend to encourage the demand for natural gas. In the absence of substantial indigenous supplies this demand can only be met by further imports of l.n.g., supplemented by high-energy s.n.g. from, initially, oil feedstocks, and possibly by the 1980s from coal. The prospects for natural gas in the longer term are thus more likely to be constrained by supply availability than by demand limitations.

What sources of supply are likely to be developed to meet this increasing demand for gas? While the delivered price of l.n.g. imports will always of course be an important factor, the continued rising prices of alternative competitive energy forms such as low sulphur oil are steadily increasing the economic radius of l.n.g. schemes. For supply security reasons, Japan will also probably wish to draw its future gas supplies from the widest possible variety of geographic and political sources, even if this incurs some economic disadvantage. The most probable supply sources at the present time (in no special order of timing or of exportable potential) are Malaysia, Indonesia, the U.S.S.R., and various countries in the Middle East (e.g. Abu Dhabi, Iran, Qatar, etc.). Australia must be excluded from this list because of the present government's policy of non-export. Figure 8 shows the geographic relationship of these potential supply sources, and some others such as Bangladesh, to Japan and also to the U.S. west coast, which could be an alternative outlet for some of these supplies.

It is clear that, bearing in mind the 'cost of distance', the Middle East is less favourably placed than other potential sources, but for the reasons mentioned above this prolific gas-bearing region is nevertheless likely to become a substantial supplier of gas to Japan. A combined l.n.g. gas liquids scheme from offshore Abu Dhabi has already reached the firm contract stage and others will no doubt follow.

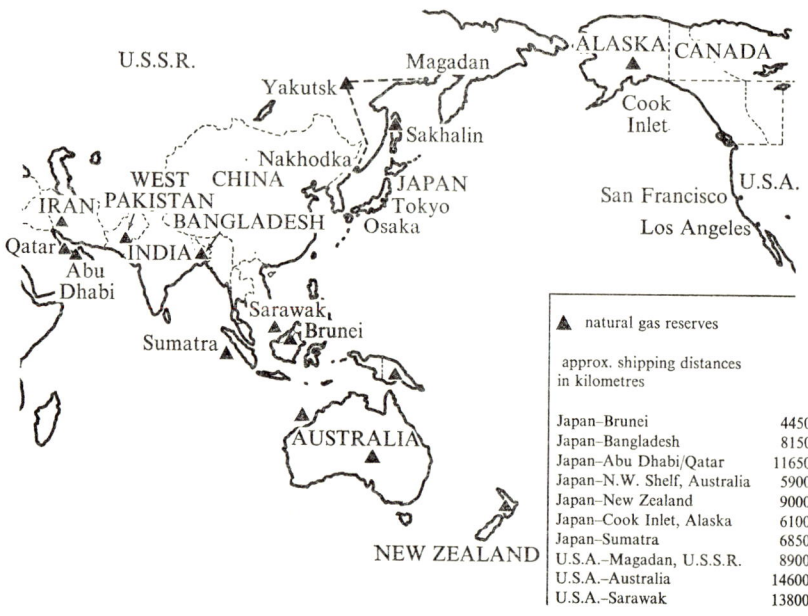

FIGURE 8. Principal natural gas reserves in relation to Japan and U.S. west coast. (1 km ≈ 0.54 nautical miles.)

Based on present technology, expected capital and operating costs, distance from market, etc., and allowing for the effects of inflation, the laid-down price in Japan of l.n.g. by the early 1980s could well be in the U.S.¢114–142/10^9 J range (¢120–150/10^6 Btu). This price range is not thought to be unrealistic or unduly out of line with the price of alternative forms of clean energy that are likely to prevail at that time.

By the mid-1980s Japan is likely to be importing from between 30×10^9 and 60×10^9 m³/year of gas in the form of l.n.g., the exact level depending on how quickly the market adjusts itself to absorb l.n.g. and the rate at which new projects can be negotiated, built and brought on-stream. To achieve this range of imports will require a very substantial effort by the industry and the governments concerned if it is to be realized over a time-span of barely 10 years. However, even at the higher level of l.n.g. imports postulated, natural gas would be supplying only about 5% of Japan's total energy demand in the mid-1980s. Like the U.S.A., but unlike the U.S.S.R., the growth for gas in Japan will be constrained by supply availability rather than demand potential in the foreseeable future.

8. WESTERN EUROPE

Figure 9 gives a broad impression of the relative size and location of the main gas reserves in Western Europe, as published by the *Oil and gas journal* (December 1972). It shows that the major deposits so far found in Europe have been in the Netherlands and in the North Sea, to which should be added the smaller but nevertheless important reserves in West Germany,

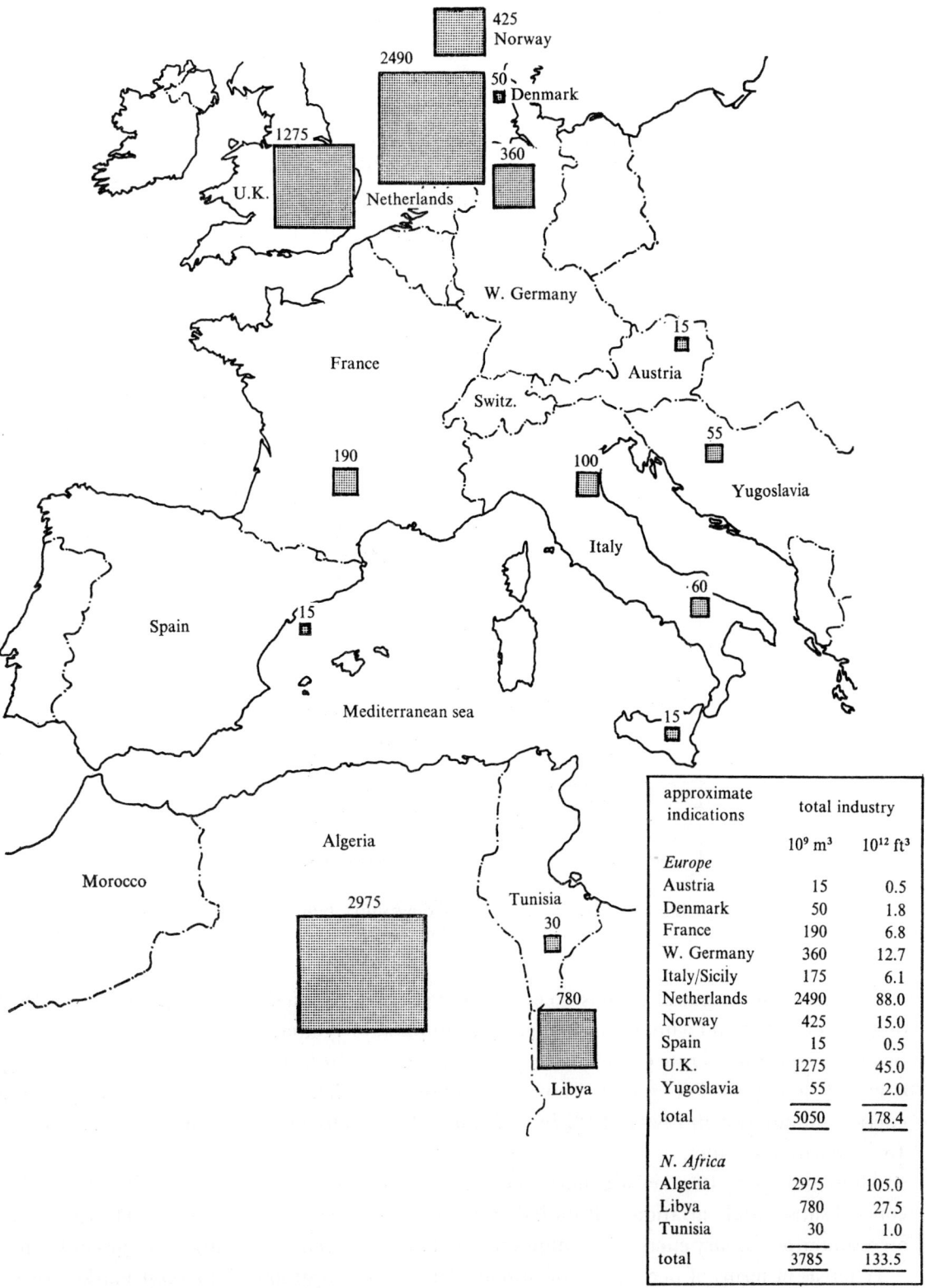

approximate indications	total industry	
	$10^9 \, m^3$	$10^{12} \, ft^3$
Europe		
Austria	15	0.5
Denmark	50	1.8
France	190	6.8
W. Germany	360	12.7
Italy/Sicily	175	6.1
Netherlands	2490	88.0
Norway	425	15.0
Spain	15	0.5
U.K.	1275	45.0
Yugoslavia	55	2.0
total	5050	178.4
N. Africa		
Algeria	2975	105.0
Libya	780	27.5
Tunisia	30	1.0
total	3785	133.5

FIGURE 9. Estimated natural gas reserves in 1972. Western Europe and North Africa – $10^9 \, m^3$ (source of data, *Oil and gas journal*, December 1972).

France and Italy. Algerian reserves are also shown on figure 9 as, like Soviet gas, these reserves could have a significant role to play in Europe's future gas supply picture. The figures for Norwegian reserves shown are in fact almost certainly now on the low side, bearing in mind recent discoveries.

The historical development of the natural gas market in Europe is well documented and it is not proposed to give a detailed account in this paper. The European gas industry is now fully committed to natural gas with conversion from manufactured gas to natural gas complete or virtually complete in the Netherlands, Belgium, France and Italy and well advanced in the U.K. and West Germany.

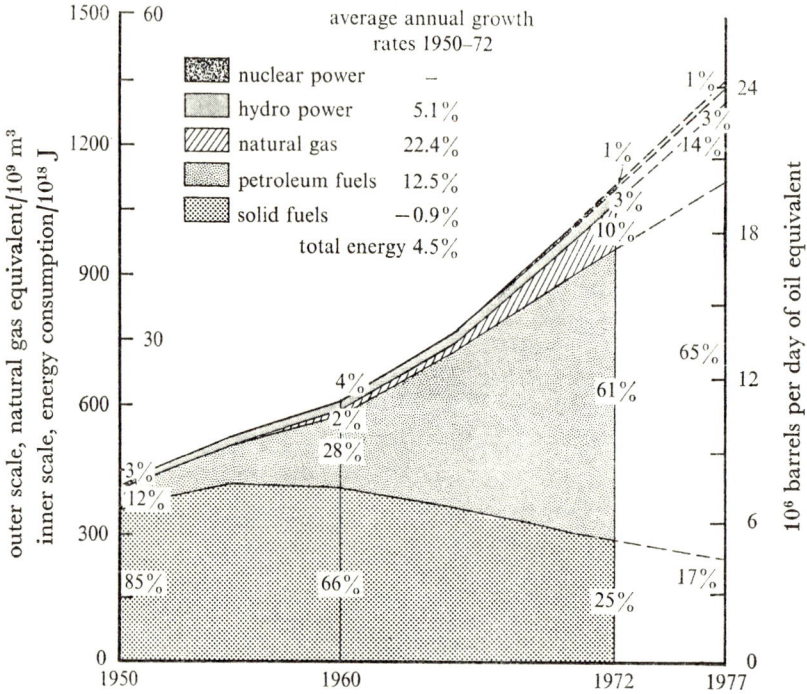

FIGURE 10. Western Europe – inland energy (excluding ship's bunkers and international aviation) consumption and forecast 1950–77.

Consumption of natural gas has exceeded most previous predictions. Between 1969 and 1972, despite generally mild winters, natural gas consumption in Western Europe as a whole increased by an average annual rate of 30 % compared with only a 5–6 % annual increase for total energy over the same period. Figure 10 shows how natural gas has grown from about 2 % of total energy consumption in 1960 to 10 % by 1972 and is forecast to increase to 14 % by 1977 reaching 15 % shortly thereafter.

Although figure 10 gives a general indication of energy developments in Europe, it does conceal substantial variations within individual countries, as is shown by figure 11, which gives an estimate of primary energy consumption in 1977 in the seven main markets for gas in Europe.

Whereas in terms of percentage of primary energy consumption, the largest market for gas will be the Netherlands, where gas is expected to supply some 60 % of primary energy consumption in the late 1970s, West Germany with about 16 % for gas and the U.K. with around 18 % are both larger gas markets in terms of quantity.

Existing indigenous reserves supplemented by established and firmly contracted imports of pipeline gas from the U.S.S.R. and l.n.g. from Algeria will, as already indicated, enable natural gas to supply about 15% of West European energy by the late 1970s, by which time all the presently known main gas fields are expected to be on or near plateau production levels. In the absence of major new discoveries and/or further imports, natural gas will inevitably steadily decline thereafter in the contribution it can make to total energy supply. While this is the overall situation for Europe in the longer term, some limited gas supply shortages are already beginning to be felt in certain localized areas in West Germany, France, Italy and Belgium; by the late 1970s a similar situation can be expected for the U.K. Some of these immediate supply deficits will be filled by gas from recent discoveries in the North Sea for which contracts have been concluded but where the pipeline system from gas fields to the mainland has yet to be built (e.g. Ekofisk and Frigg) or in some cases which have yet to be firmly committed (e.g. Brent). Also some additional imports of l.n.g. from Algeria are currently being negotiated. However, these new identifiable supplies are only likely to push the supply shortage back a few years.

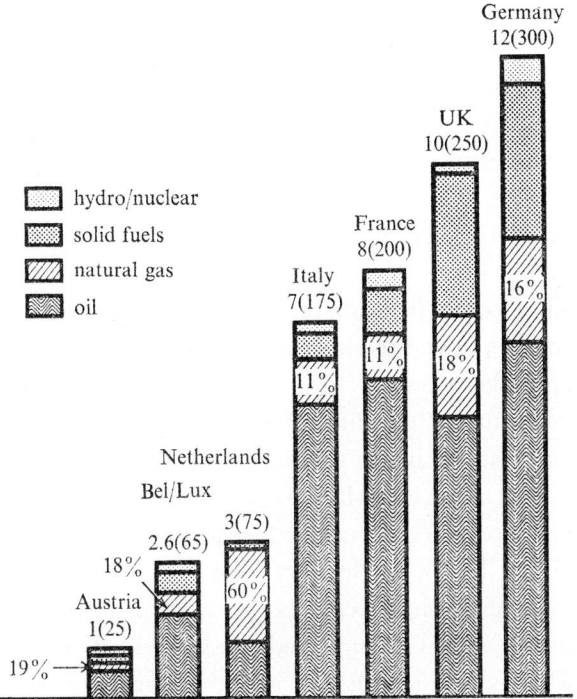

FIGURE 11. Estimated primary energy consumption in 1977 – 10^{18} J (10^9 m^3 of natural gas equivalent).

What then is the future for gas in Europe? Will total gas supply level off in the late 1970s or can the gas industry continue to grow? The market pressures for growth in gas, already strong will no doubt intensify. By the late 1970s virtually all Europe's gas distribution systems will have been converted from manufactured gas to natural gas. Other energy forms, in particular oil, are expected to be in short supply. Environmental pressures, which favour the use of a clean fuel such as natural gas, will increase. Increasing prosperity and the concomitant desire for better living standards will enhance the demand for gas for space heating and similar 'desirable' applications. All these and other related factors will raise the demand for gas within the present inter-fuel price relationship. In an unrestricted supply situation it is entirely reasonable to suppose that natural gas could take up a third or more of Europe's energy demand. But can it in practice? The answer is, most probably, no, because even under the most optimistic assumptions the market is likely to remain supply-restricted.

What are the prospects for new supplies? First and foremost will be the development of such new reserves as may be discovered in and around Europe. The European offshore, in particular, is still a relatively new petroleum exploration area. Most of the early discoveries were of gas and now the pendulum is swinging towards oil, but nobody can forecast with any degree of precision what new exploration will offer in future – oil, gas, both or nothing. All one can conjecture is that with many new areas such as the Western Approaches, the Irish and Celtic seas, and large parts of the North Sea, the Adriatic and the Mediterranean yet to be explored, surely more gas reserves will become available in the years ahead. However, it would take an optimist to assume that the industry will discover the equivalent of an additional Groningen or several Leman Bank gas fields which would be necessary if natural gas were to achieve a substantial increase in its share of future European energy consumption.

Secondly, further imports of pipeline gas from the U.S.S.R. and l.n.g. imports from Algeria are probable. Although as already discussed the potential reserves in the U.S.S.R. in particular are very considerable, most of the readily exploitable reserves in these two countries appear to be already committed to various markets and it would seem that a breathing space is now likely in which they can build facilities to fulfil their commitments and appraise their undeveloped potential reserves before embarking on a 'second round' of export commitments. Most probably they will have more gas to offer in the years ahead but it appears unlikely that this second phase of commitment could make any worth-while contribution to Europe's gas supply picture much before the 1980s.

Thirdly, s.n.g. from oil or coal will probably have a role to play, but European coal is not cheap and oil prices are moving upwards. s.n.g. is likely to be very expensive and possibly limited in the contribution it can make.

Fourthly, l.n.g. imports from farther afield than Algeria, such as from West Africa and the Middle East, may well be arranged in the years ahead, but, like s.n.g., such l.n.g. supplies will be expensive as a consequence of the long shipping distances and consequent high freight costs.

Lastly, there is the possibility that gas may be brought to Europe by large-diameter pipeline from the Middle East, either by direct delivery – for example, through Turkey, Greece and Yugoslavia – or by exchange as in the recently announced project between Iran, the U.S.S.R. and West Germany. The cost would be high and there are likely to be political problems, but nevertheless such schemes may well play a part in the European supply pattern in the 1980s.

The complexities of these various new and supplementary sources of gas are such that any long-term forecast becomes so highly speculative and wide-ranging as to have little practical meaning. Nevertheless, and in spite of these reservations, sufficient new supplies should be forthcoming to maintain some growth for the European gas business through the 1980s and perhaps for some time thereafter, although probably not to the extent that all market opportunities for gas can be taken up in all the main European markets of interest. This supply constraint will mean that gas distribution companies in Europe, as now in the U.S.A., will become increasingly selective and will concentrate supplies upon those outlets (domestic and premium industrial) which can support higher cost gas and where gas has special economic, social and environmental advantages over other fuels.

9. OTHER POSSIBLE SUPPLY SYSTEMS

So far the paper has discussed the role that s.n.g. from coal or oil and l.n.g. imports may play in the future in supplementing indigenous supplies of natural gas. S.n.g. from oil and l.n.g. are proven and s.n.g. from coal is confidently expected to begin to make its contribution by the end of this decade. What other possibilities are in prospect?

Appreciable publicity has been given of late to the conversion of natural gas to methanol (or methyl fuel), which as a liquid product could be shipped much more cheaply than l.n.g. In the receiving country the methanol could either be re-formed to pipeline-quality natural gas, or it could be used directly as a low-pollutant fuel for industrial application. Manufacturing methanol from natural gas is substantially more expensive than the gas liquefaction process but, because cheaper ships can be used, the methanol conversion route may offer an overall economic advantage over an l.n.g. system in cases where very long ocean transportation distances are involved – for example, from the Middle East to the United States. With the increasing attention that is being paid to the exploitation of natural gas from such distant sources, some commercial projects for manufacturing and shipping methanol are being actively investigated and may materialize in the years ahead.

An approach which attracts by its theoretical elegance but which has not yet found practical application is that of the 'cold carrier' system. Work has been carried out upon various such systems, in which the cold arising from the re-gasification of l.n.g. in the country of receipt is transported back to the gas-producing country in order to reduce liquefaction costs. If such a system were to prove competitive with conventional l.n.g. shipments, this would be particularly over short ocean distances (for example, across the Mediterranean). With further technological development, and improved economics, such systems may yet have a role to play.

In the past a number of gas transportation systems using different combinations of low temperatures and high pressures have been studied, but so far nothing has been found to compete with the by-now conventional l.n.g. system, in which the natural gas is kept at or below its boiling point at virtually ambient pressure.

Very large rigid airships each capable of transporting, say, 3×10^6 m^3 (1×10^8 ft^3) of natural gas in the gaseous state are another possible method of transport. This idea is at present in the preliminary evaluation stage but may in time prove a viable alternative to l.n.g., particularly over shorter distances (say, less than 3200 km) and in this event would bring a new flexibility to gas transportation. Because of the necessary development time, it is improbable that such airships could become a commercial proposition before the mid-1980s.

The profitable utilization of the cold released when l.n.g. is re-gasified in the reception terminal is an area which offers obvious scope for improving the economies of l.n.g. schemes. Although a variety of technically valid applications for this cold exist, e.g. ethylene manufacture, the food industry, air separation, desalination, etc., such applications do not make a substantial financial contribution in any of the existing or announced l.n.g. projects. The reason for this seems to be commercial complexity rather than technical difficulty; as more l.n.g. projects come into operation it is certain that schemes to make use of the cold available at the receiving terminals will be developed.

Finally, a long-term prospect which does not involve natural gas as such but which may be considered to fall within the scope of this paper since it would make use of existing natural-gas transportation and distribution pipeline systems, is the so-called 'hydrogen economy' concept, in which hydrogen produced with the aid of nuclear energy (either via electricity generation followed by the electrolysis of water, or by a combined nuclear heat/chemical process) is fed into a pipeline system and distributed for the same sort of applications as natural gas. The interest aroused by this concept is due to the fact that it offers an attractive alternative to total electrification of the world's economy at that time in the future when nuclear power generation becomes the principal source of energy. Advantages claimed for it, compared with electricity, include:

(i) transmission and distribution of energy more efficiently and at lower cost;

(ii) less disfigurement of the environment;

(iii) easier storage;

(iv) potential for use as a hydrocarbon fuel component, as a chemical feedstock and as a feedstock for fuel cells.

However, a development along these lines is unlikely to become a significant factor in world energy supplies until the cheaper recoverable fossil fuels are nearing exhaustion and nuclear generating capacity is available on a massive scale. It therefore almost certainly falls outside the period under discussion in this forum.

10. CONCLUSION

World resources of natural gas are theoretically more than sufficient to match world demand. However, for economic, geographical, logistic, political and other reasons, supply shortages are already apparent in the United States and are beginning to emerge in Western Europe.

Gas which can be transported to market by pipeline is, and will remain, the mainstay for the continuing development of most natural gas markets. L.n.g. imports and s.n.g. are, or will become, useful and important sources of supply, but their role will usually be to supplement and not to replace pipeline gas. The notable exception to this is Japan, where l.n.g. imports, in the absence so far of any indications of worth-while indigenous energy supplies, will almost certainly provide the base source of supply for future development of the Japanese gas market.

'Cost of distance', in spite of a growing world-wide appreciation of an energy shortage, is likely to remain an inhibiting factor in matching potential sources of gas with potential demand. Less immediately apparent but possibly of even greater inhibiting effect is the fact that, even where the theoretical economics of an international l.n.g. project appear technically and financially viable, the complex problem of reaching agreement on the intricate political, fiscal and commercial aspects of the venture (which may involve capital expenditure well in excess of U.S. $\$1 \times 10^9$) may be such as to outweigh all else as a factor for delay, possibly causing years to be lost or even (in extreme cases) the complete missing of a market opportunity.

In a number of markets natural gas prices have been either arbitrarily regulated or otherwise linked to the cost of relatively low-value alternative energy forms. This has resulted in the expansion of the use of gas for inferior purposes such as steam-raising, where other fuels would have been more suitable. Prices for natural gas in such markets will probably move upwards in the future to equate with the true market value appropriate for this clean, versatile and flexible form of energy. This will help to ensure that natural gas is economically employed in those applications and market sectors where its inherent qualities can be utilized to the maximum advantage, and at the same time will provide added encouragement and incentive for the search for and development of new prospective gas-bearing areas.

The task which lies before the gas industry is:

(1) To press ahead with the full and proper exploitation in the market-place of existing and prospective gas reserves located in or close to the main markets.

(2) To attack the problem of linking distant gas sources to markets by continuing the development of supply systems based on l.n.g., large-diameter pipelines or whatever other methods may prove competitive.

(3) To establish and foster with governments a clear understanding of the problems and opportunities so that complex projects based upon such systems may come into operation without undue delay.

(4) To sponsor research into and development of all other economic forms of supplementary gas supplies such as synthetic natural gas from oil and coal, conversion of natural gas to methanol (where this offers advantages over l.n.g.) and any other new processes or systems which can contribute to the supply pattern.

In this manner the distinctively valuable contribution made by gas to meeting the world's ever-expanding need for energy can be maintained and reinforced.

Discussion

DR W. C. MARSHALL, F.R.S. (*A.E.R.E. Harwell, Didcot, Berkshire*)

Once we have the pipelines and infrastructure of a gas industry, I can see the argument for converting coal to 'synthetic natural gas', but in the long run it is not wiser to think primarily in terms of converting coal to a synthetic crude oil, because the latter process would make less demands upon a hydrogen supply?

Mr COPPACK's reply was to agree with this comment but to stress that the infrastructure of a gas distribution system would, in practice, argue strongly for a significant conversion of coal to synthetic natural gas indefinitely.

Phil. Trans. R. Soc. Lond. A. **276**, 485–493 (1974)

Printed in Great Britain

Hydro (including tidal) energy

By K. R. Vernon

North of Scotland Hydro-Electric Board, Edinburgh

This paper reviews, on a world-wide basis, the production from hydro-electric installations and forecasts which have been made of potential water-power resources. Various types of development are described. It considers how hydro-electric resources can best be used in an interconnected electricity supply system with thermal and nuclear generation, including pumping to assist in the economic use of other fuels. Hydro-electric development in the United Kingdom is briefly described.

   The availability of tidal energy is reviewed, and reference made to the Rance Tidal Power Station in France.

1. Hydro

1.1. *Production*

The United Nations' survey for 1970 shows that the world production of hydro-electric power was 4.24×10^{18} J (1178 terawatt hours) and formed 24 % of the total production of electricity from all sources.

Put in a more domestic context, this production was equivalent to 4.7 times the total electricity generated in the U.K. that year. For those who prefer to work in coal equivalent – that is, if generated instead from coal-fired plant – hydro power represented a saving of 600×10^6 tonnes.

Table 1 shows the production in 1961 and 1970 in the developed countries, centrally planned economies and developing countries. It is significant that while in 1970 the developed countries produced almost three-quarters of the world's hydro energy, the rate of development of hydro elsewhere over the past decade has been twice as fast. This trend will become more marked with the approaching exhaustion of hydro potential in the developed countries.

Table 1. World hydro production

	1961		1970		increase
	10^{18} J	TW h	10^{18} J	TW h	(%)
developed countries	2.07	575	3.05	847	47
centrally planned economies	0.32	89	0.67	185	107
developing countries	0.22	61	0.52	146	137
total	2.61	725	4.24	1178	62

1.2. *Hydro resources and their use*

Any review of estimates of potential hydro development has to be treated with considerable caution. The figure of 82.8×10^{18} J (23 000 TW h) given at the 1968 World Energy Conference (that is, twenty times 1970 production) must in my view be regarded as more a technical than economic assessment and is probably optimistic by a factor of between 2 to 3. On this basis a more realistic assessment might be as shown in table 2. The pace and manner in which these remaining resources are exploited will be largely governed by the stage of industrial development of the country in which they are located.

TABLE 2. WORLD HYDRO RESOURCES

	potential		developed, 1970
	10^{18} J	TW h	(%)
developed countries	4.68	1300	65
centrally planned economies	6.47	1800	10
developing countries	18.0	5000	3
total	29.15	8100	14.5

The maximum use of the available head of a typical river system can lead to a wide variety of types of installation. These can range from stations with large reservoir capacity located in the upper reaches of the catchment to less well regulated stations farther downstream, and finally, where the fall in the river is small but the flow large, a series of run-of-river stations each with the generating plant integral with a barrage constructed to create sufficient usable head.

The history of hydro-electric development in those countries initially with abundant hydro power shows that the first phase is usually the provision of dependable power from key stations with large reservoir capacity, followed only in the more final stages by the construction of run-of-river stations whose variable output can by then be more readily absorbed into the system.

1.3. *Developed countries*

The vast bulk of the remaining resources in developed countries now lies in Canada and this is increasingly being exploited, for use also in the U.S.A., by the use of long-distance transmission.

In Europe some 80% of the hydro potential is expected to be exploited by 1980. While there are isolated developments taking place in many countries the most significant work under construction is a series of run-of-river stations on the Rhone and Danube, which will also form part of an improved navigational scheme. There is also a trend to augment the output of existing schemes by the diversion of additional catchments.

In the U.K., and in Scotland in particular, virtually all the hydro resources have been developed. There are a few possible schemes but these are not economic at today's interest rates and too small to justify promotional problems.

In the U.S.A., Japan and Australia a high proportion of the resources have been exploited. Those schemes in hand tend to be multi-purpose, covering, as in Europe, irrigation, flood control, water supply and navigation.

1.4. *Centrally planned economies*

The major developments in Russia lie principally in Central Siberia, where, for example, some six stations in the 4000–6000 MW range are under construction or planned on the Angara and Yenisei rivers. The output of these and existing stations will be used initially to meet industrial expansion in the region. Other schemes in hand are in the Caucasus and in Central Asia. While the Russians are developing high voltage d.c. transmission towards Central Asia no plans extending these to Siberia to make use of the vast hydro resources there have been published.

Little is known about hydro development in China, where there are vast resources available. No doubt these will be developed to match industrialization, with the added multi-purpose incentives of flood control, irrigation and navigation.

1.5. *Developing countries*

As indicated earlier, more than half the world resources are in the developing countries and a negligible proportion has been harnessed. Only a limited number of sites have been studied closely but it is evident that, among many other countries, Africa and South America in particular have very substantial potential.

It is unfortunate that at present where hydro is abundant the demand for it is small and that meantime energy, at a time of shortage and high cost elsewhere, is flowing to waste. However, much of the world's long term anticipated growth in energy needs is linked with increase in population and industrialization of the developing countries, and many of these countries are well endowed with hydro. It would seem that economic and political factors will encourage such countries to maximize the use of these indigenous hydro resources to meet much of their growing energy needs rather than compete unnecessarily for world-wide reserves of fossil fuel. Similarly, high energy prices will probably encourage the greater use of transmission as a means of utilizing surplus hydro, and there will almost certainly be an increasing trend back towards the location of electrically intensive industries in these hydro-abundant countries.

2. PUMPED STORAGE

2.1. *World-wide scene*

It may appear inappropriate in a discussion on energy resources to refer to pumped storage – a device which consumes more energy than it produces. On the other hand, the introduction of pumped storage into a thermal generating system enables that system to be operated more efficiently with overall economy of fuel. Furthermore, it offers an additional means of using nuclear fuel to replace fossil fuel.

It is not surprising, therefore, that with the growing use of large base load thermal sets there has been increasing interest in peak load plant in the form of pumped storage. The progressive exhaustion of conventional hydro-electric sites in developed countries has been a further incentive to seek out plant with similar flexible operating characteristics.

For these reasons, while the first use of pumped storage for electricity supply goes back to the 1930s in Germany, its real development took place after the war, as can be seen in figure 1. As a result, the pre-war capacity of 1000 MW rose to 20 000 MW in operation by the end of 1972, half in Europe and the balance mainly in the U.S.A. and Japan. A similar amount of plant is under construction.

The first major installation in the British Isles was in the early 1960s with the 360 MW scheme at Ffestiniog in North Wales, and was followed in the mid-1960s with a 400 MW station at Cruachan in Scotland. Under construction are two stations of about 300 MW, one at Foyers in Scotland and the other at Turlough Hill in Southern Ireland. The Generating Board are at present seeking Parliamentary approval for a further scheme in North Wales with a capacity of 1440 MW.

2.2. *Siting and plant considerations*

The basic requirements for an economic pumped-storage development are the existence of upper and lower reservoir sites in close proximity with a difference in level of at least 150 m. The site should be within reasonable distance of centres of load and off-peak generation.

A feature of many overseas schemes is the creation of hilltop reservoirs. With a more natural reservoir site the opportunity is usually taken to collect any run-off from the adjacent catchments so as to improve the overall economics of the scheme. Indeed, several pumped-storage installations are developments of conventional hydro-electric sites where the natural flows were too small to give a viable scheme.

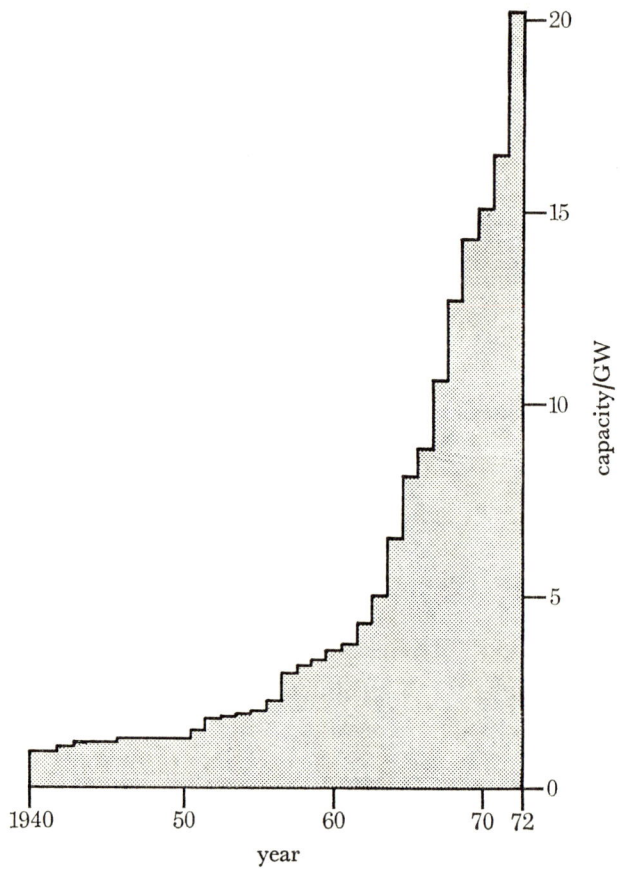

FIGURE 1. World pumped storage capacity (in operation).

Good sites, such as that at Cruachan, are unfortunately hard to obtain. Coastal cliff sites offer possibilities but require precautions to prevent percolation from the upper reservoir into the local freshwater supplies. A more recent idea is the underground excavation of the lower reservoir with the upper reservoir located at ground level.

2.3. *Use on the system*

To the system operator the attraction of pumped storage lies in its dual role of very fast response peak load generating capacity to replace loss of plant elsewhere, and a pumping load which can be switched on at night to avoid shutting down high-merit thermal plant. As a consequence, significant system fuel savings result from the more planned and level operation of thermal plants, including a reduction of spinning reserve, hot standby and other duties. Added to these are the fuel savings resulting from the replacement of peak load generation from low-merit thermal plants with energy pumped at night from high-merit plant. The case for pumped storage, both on economic and fossil fuel conservation grounds, becomes greater when the amount of nuclear power on the system exceeds the night load and becomes available for pumping, thus effectively providing peak load power from nuclear energy.

There are limits, however, to the amount of pumped storage which can be used on a system since there is a finite amount of low duration peak load. Pumping is normally only economic at night and weekends, and if demands of longer than 6 h duration are to be met this can only be achieved by additional upper reservoir capacity or by increasing the pump capacity relative to the turbine capacity. The economic penalty of increasing pumping capacity may be high but could be justified in the future with a large nuclear programme.

FIGURE 2. Principal tidal power sites.

3. TIDAL ENERGY

3.1. *Sites and potential*

Tidal energy was used in the Middle Ages in small tide-mills along the Atlantic coasts of Britain, France and Spain, but as other power sources became available they fell out of use. However, there has been continuing interest in the development of large scale tidal power schemes for the generation of electricity. The places where high tidal ranges occur are shown on figure 2, interest having been centred on Brittany, the Severn, the Bay of Fundy in North America and the White Sea and the Sea of Okhotsk in Russia.

High tidal ranges occur due to the funnelling effect of particular shallow coastal estuaries and gulfs. The construction of a tidal barrage may increase or reduce the tidal range, depending on the resonance characteristics of the estuary. The other important parameter is the volume of the basin impounded by the tidal barrage, as this, together with the vertical range of tides, dictates the capacity of the site.

It has been estimated that the total output of sites capable of practical development is of the order of 1.26×10^{18} J (350 TW h) per annum. As this is only one and a half times the consumption of electrical energy in the U.K. it is clear that the potential contribution by tidal energy to world energy requirements is limited.

3.2. *Development problems*

The main drawbacks to development are the cost, the lunar rather than solar cycle of the tides, and the amplitude range of about 3 to 1 from spring to neap tides. As a consequence, generation at a development with a simple single basin scheme operating on the ebb-tide rarely coincides with the demands on the electrical supply system; such schemes do not save investment on other power stations and can only be credited with fuel savings. Complicated multiple-basin schemes to re-time some of the energy so as to produce some firm power would be much more costly.

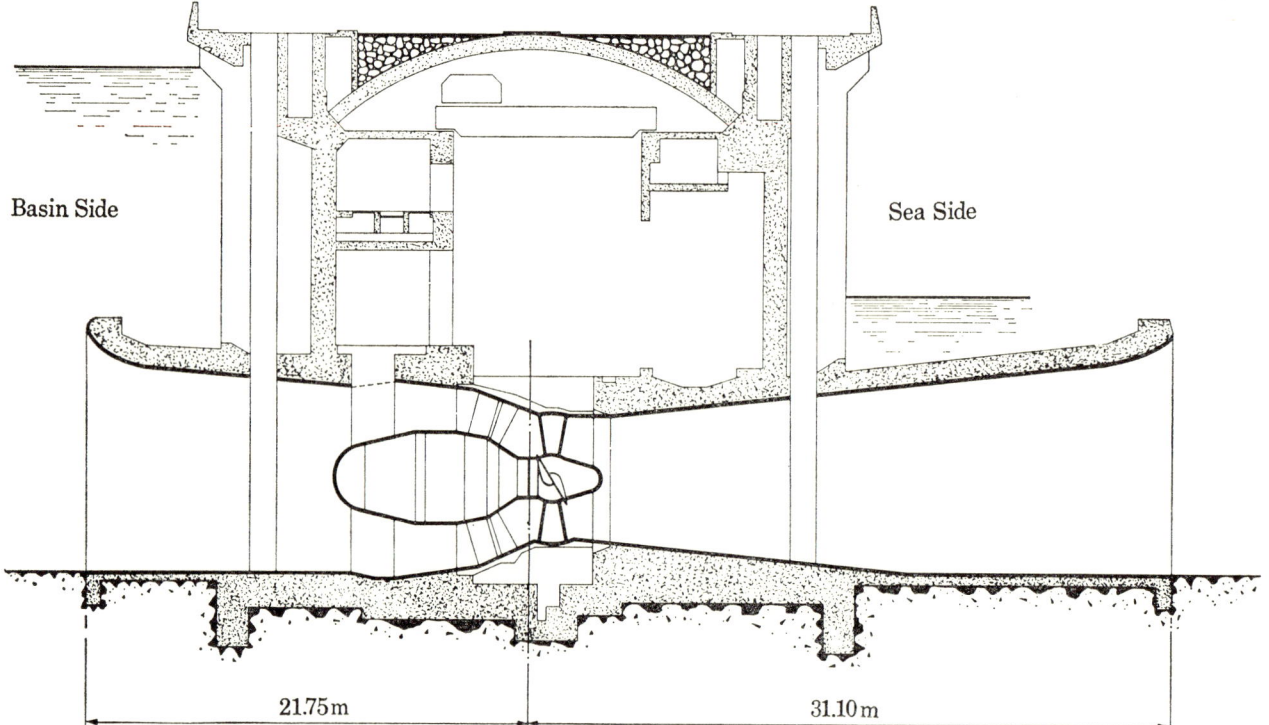

Basin Side   Sea Side

21.75 m   31.10 m

FIGURE 3. Rance tidal power station.

3.3. *Rance scheme*

The development of the horizontal reversible bulb turbine by the French, as shown in figure 3, resulted in the building of the 240 MW Rance single-basin scheme in 1966 to generate with flow in either direction; in addition, the machines can pump to increase the effective head range or ensure stored water, allowing some credit to be given for firm power. The scheme, believed to have been very expensive, was to have been the forerunner of a large scale double-basin development on the coast of Brittany at Iles de Chausey with a capacity of 3000 MW and an output of 46×10^{15} J (12.8 TW h).

3.4. *Russia*

The Rance Station was built in the dry between cofferdams, but current thinking, in order to save construction cost, favours the use of large prefabricated units which would be floated to site and sunk on prepared foundations. This method was adopted by the Russians at the small Kislaya Station north of Murmansk, which was commissioned in 1968 as a forerunner of the

2000 MW Mezen Bay project on the White Sea. It is reported that feasibility studies on this project were started in 1971 and that a vast project (25 000 MW) on the Sea of Okhotsk is under consideration.

3.5. *British Isles*

In the British Isles, where there are a few favourable sites on both coasts of the Irish Sea, the Severn Scheme was studied by a government commission which reported in 1945. Judged against the saving in coal which would result from the annual production of 8×10^{15} J (2.2 TW h), this 800 MW scheme was considered to be uneconomic at the prevailing low interest rate of 3 % per annum. Since then there have been significant advances in the development of low head turbines and in civil engineering techniques and a much larger scheme has been envisaged. Nevertheless, the current higher level of interest rates must make any such scheme unattractive unless the costs were shared in a multi-purpose development incorporating improved road and harbour facilities.

3.6. *North America*

On the other side of the Atlantic there have been a succession of investigations into tidal developments within the Bay of Fundy involving both Canada and the U.S.A. Out of these have come four separate schemes, totalling about 5000 MW of capacity with a production of 53×10^{15} J (14.8 TW h) of energy. Recent investigations have shown that developments would only be economic at an interest rate of not more than 4 %. However, fresh interest into these projects has led to the setting up of a Tidal Power Corporation in Nova Scotia to carry out further investigations.

3.7. *Summary*

While much effort and ingenuity has been devoted to tidal power, an economic solution has yet to be demonstrated. It has to be accepted that with the present technology, the low concentration of power in the sea, even at the most attractive sites, must lead to physically large and thus expensive works; also that the energy produced is rarely in phase with man's needs. These present a formidable challenge.

Discussion

Mr P. WILSON (*Gilbert Gilkes & Gordon Ltd, Kendal, Westmorland*)

Mr Wilson said that he was the Chairman of a firm of water-turbine manufacturers and that he was not going to ask Mr Vernon a question but that the point which he wished to make could have been a question to the Secretary of State. He would have liked to have asked Mr Peter Walker if the government would immediately instruct the River Authorities to stop making a levy on all in England and Wales who generated electricity by water power.

In accordance with clause 58 of the Water Resources Act 1963 the River Authorities had powers to make a charge for river water which they claimed was 'abstracted' when used for power generation. As a result many small hydro-electric units had been closed down and there was little likelihood that any more would be installed. No attempt had been made to introduce such legislation in Scotland, where there would have been an immediate outcry.

The generation of hydro-electric power did not use up coal, oil, nuclear material or any of the world's limited resources and should, therefore, be encouraged to the full. He was certain that it might not be long before the owner of quite a small hydro-electric plant might be greatly envied by his neighbours. In his opinion it was ridiculous to make a charge on those who used water power when there was likely to be a fuel crisis and when, in any case, it was acknowledged that there would soon be a big increase in the price of all fossil fuels.

K. R. Vernon

I was surprised to learn from Mr Wilson that small hydro-electric stations in England and Wales are charged for abstraction of water, particularly as these small schemes must return the water almost directly and unpolluted to the river concerned. One justification of a charge, however, would be the improvement of dry weather flows in the river consequent upon the provision of a reservoir storage capacity upstream.

From enquiries received at my office I known there is a real interest throughout the country in small sites with hydro-electrical potential, and it would be a pity if unjustifiable levies prevented their development.

S. H. Salter (*Bionics Research Laboratory, School of Artificial Intelligence, University of Edinburgh*)

Power from sea waves?

The energy available from a wave can be calculated from the change of potential energy as the water above sea level falls down into the trough. This can provide

$$\text{power} = kw\rho A^2 g \sqrt{(g\lambda)},$$

where w is the width, ρ the density of water, A the amplitude, g the acceleration of gravity, λ the wavelength, and k a constant $= \frac{1}{6}\sqrt{(2\pi)}$ for triangular waves.

Observations of waves in the North Atlantic show that power levels between 10 and 100 kW/m may be expected. I have been testing wave-tank models in which more than 50 % of the energy is extracted. There are obvious problems concerned with storage, transmission and maintenance, but if the technique can be extended to full scale then the electrical requirement of the U.K. could be satisfied by a few hundred kilometres of installation floating out at sea off the west coast of Scotland. (See Draper, L. & Squire, E. M., 1967, Waves at Ocean Weather Ship *India* (59° N, 19 W°). *Trans. R. Inst. Nav. Arch.* **109**, 85–93.)

K. R. Vernon

Mr Salter drew attention to the potential power of sea waves. In connexion with tidal projects I mentioned the formidable challenge of harnessing the low concentration of power in the tides. This applies equally to utilization of wave power, which would have the same disadvantages as tidal power of not always being available when required and not saving investment in other power stations. It is interesting nevertheless to learn Mr Salter has managed to extract more than 50 % of the theoretical energy in wave tank tests.

Mr N. E. Butcher (*Open University in Scotland, 60 Melville Street, Edinburgh*)

I wonder if Mr Vernon would comment on the possible contribution that could be made to our energy requirements by windmills linked to pumped water storage, since there would appear to be no great shortage of wind in Britain.

K. R. Vernon

The same basic problem of intermittent output applies to windmills so that the value of their output is limited to fuel savings. Linking windmills to pumped storage schemes would not overcome this inherent economical disadvantage. On the practical side my Board cooperated with the Electrical Research Association in the construction and operation of a 100 kW wind-power plant in Orkney in the early 1950s. Severe problems were experienced with the stability both of the structure, which was 24 m high, and of the rotor which was 15 m in diameter. In addition, there were difficulties with the control of the feathering blades. Despite modifications to the plant it was concluded then and our opinion still is that large windmills were not practical or economical for public electricity supply.

Phil. Trans. R. Soc. Lond. A. **276**, 495–505 (1974)
Printed in Great Britain

Natural sources of nuclear fuel

By S. H. U. Bowie

Geochemical Division, Institute of Geological Sciences, London

Reserves, resources, potential resources (or potential) are defined and related to price categories and it is deduced that sufficient uranium is available from reserves (866 000 tonnes U) at a price of less than $20/kg U_3O_8 to meet estimated demands to the end of the 1970s. However, this will mean rehabilitating mines that have shut down and commissioning new or regenerating old refining plant.

There is a risk of a shortage of uranium by the mid 1980s if safeguards are not taken in time to ensure that uranium can be made available from reserves and estimated additional resources (916 000 t U) in the < $20/kg U_3O_8 price category presently indicated as being in the ground. Resources in the price range $20–30/kg U_3O_8 are over 10^6 t U and in the $30–60/kg U_3O_8 are of the same order or more.

It is considered that prospecting on a world-wide basis should be increased appreciably and maintained at a reasonable level. Also, that new search techniques should be developed and applied mainly to the discovery of hidden ore-bodies. If this is done in good time, there is no reason to predict any absolute shortage of relatively low-cost uranium this century.

Thorium has a potential in high-temperature gas-cooled reactors. Currently known resources are of the order of 10^6 t ThO_2.

1. Introduction

The difference between reserves, resources and the abundance of uranium and thorium in the upper part of the Earth's crust is not generally appreciated. This has resulted in widely differing estimates of tonnages likely to be available as fuel for nuclear reactors. 'Ore' is defined as material that can be mined, beneficiated and sold at a profit. This implies that 'ore reserves' refers to material in known ore-bodies that is likely to be commercially viable at some time in the near future. 'Resources' refers to all material likely to be available at a realistic price, but which is unlikely to be exploited until reserves have been largely depleted.

At present there is no recognized market price for uranium, but, in order to calculate ore reserves, it is widely accepted that material available at a price of < $20/kg U_3O_8 in concentrate form classifies as ore. Resources are categorized as material likely to be available at < $20/kg U_3O_8, at $20–30/kg U_3O_8, and at $30–60/kg U_3O_8. There are no equivalent categories for thorium reserves and resources, tonnage estimates being related to material available at a price of < $20/kg ThO_2 in concentrate form.

Ore that is deduced from geological or other evidence, but on which few if any quantitative measurements have been made, is termed 'potential ore'. There is an element of potential ore in material classified as reserves. This results because, in calculating ore reserves, geologists generally use the carefully defined categories of 'measured', 'indicated' and 'inferred' ore and in the latter category there is some overlap with potential ore. The boundaries between reserves and resources are not hard and fast and errors in calculating the amount of ore in the ground can range from a few to as much as 50 % of the actual tonnage, depending on the nature of the orebody. Relatively homogeneous deposits of uranium in flat-lying sediments are by far the easiest to assess, veins and pegmatites the most difficult. Fortunately, over 70 % of reserves of uranium are disseminated in quartz-pebble conglomerates and sandstones so that the estimates of tonnage can be accepted with a reasonable degree of confidence.

The terms 'potential resource' or simply 'potential' relate to material the location of which has not been defined but which might become available in the more distant future at a realistic price. The ultimate tonnage of uranium or thorium in the Earth's crust should neither be considered a resource nor a potential resource. Estimates such as those of Lewis (1972), who gives the figure of 10^{14} tonnes U in the Earth's crust, of which 0.25×10^{14} t is within 1.6 km of the surface, are of no value in establishing the tonnage of uranium that is likely to be utilized as nuclear fuel. Likewise, attempts to calculate ultimate uranium-ore tonnages from crustal abundance data are not helpful. What is of major importance is an understanding of the processes that resulted in uranium being concentrated in abnormal amounts in certain parts of the Earth's crust at different eras throughout the Earth's evolution.

The cut-off grade that may eventually be reached is difficult to estimate but it could easily be of the order of 50 parts/10^6 metal, which is a factor of about 10 below the lowest grade ore that could be mined at present in favourable circumstances. At this level many millions of tonnes of uranium could be made available to power breeder reactors. It is thus unnecessary, from the practical viewpoint, to attempt to estimate ultimate tonnages or to contemplate extracting uranium from rocks such as normal granites which have an average uranium content of about three times the crustal abundance of 1.6 parts/10^6.

2. Uranium distribution

Uranium is widely distributed in the Earth's crust in minerals of varying degree of chemical complexity especially in association with acid igneous rocks. The element appears to have been concentrated, along with other volatile elements, in upper crustal rocks in early Precambrian times. Thus, some 90 % of presently known ore reserves occur in well-defined provinces in Precambrian cratonic masses such as the Shield areas of Australia, Canada and South Africa or in sediments immediately overlying Precambrian rocks as in the case of the Colorado–Wyoming uranium province.

Recent geochronological studies have shown that the processes of introduction or remobilization of uranium minerals generally extended over a long interval of time. For example, in the Blind River–Elliot Lake province of Ontario discrete uranium- or thorium-bearing mineral species are from 2500 to 600 million years (Ma) old. The same, though less extensive, time interval between early and late introduction of uranium applies in the case of phanerozoic rocks. For example, in the Colorado Plateau uranium was introduced into sediments during at least two periods – one 210 Ma and the other 110 Ma ago (Miller & Kulp 1963).

Information of this type is invaluable as a guide to exploration geologists in the search for additional reserves and resources, since once a uranium province has been recognized, detailed attention can be given to the physico-chemical controls that resulted in the element being concentrated to an economic tenor. An outstanding example of the value of geochronology to the discovery of major additions to the reserves of the Colorado–Wyoming uranium province is provided by the work of Stieff, Stern & Milkey (1953), who showed that fresh uraninite from the Happy Jack Mine, White Canyon District, Utah, is 65 Ma old. Before 1953, it was widely accepted that uranium in the sandstones and mudstones was syngenetic and essentially confined to the Morrison and Entrada formations of Jurassic age (≈ 150 Ma old). The revelation that at least some uranium in the province was much younger than the host rocks led to the widening of the search to rocks of all ages in the region. The result was the discovery of major new deposits in sediments ranging in age from Triassic to Miocene ($\approx$ 220 to 20 Ma old).

(a) Economic uranium minerals

Although uranium is an important constituent of about 100 minerals, significant production of uranium concentrate has been from only six primary and five secondary species. The primary minerals are uraninite UO_2, and the less-well crystallized variant pitchblende; coffinite $U(SiO_4)_{1-x}(OH)_x$; brannerite $(U, Y, Ca, Fe, Th)_3Ti_5O_{16}$; davidite $(Fe, Ce, U)(Ti, Fe)_3$ $(O, OH)_7$; uranothorite $(Th, U)SiO_4$; and uranothorianite $(Th, U)O_2$. Secondary species are carnotite $K_2(UO_2)_2(VO_4)_2.nH_2O$; tyuyamunite $Ca(UO_2)_2(VO_4)_2.9H_2O$; torbernite $Cu(UO_2)_2(PO_4)_2.12H_2O$; uranophane $Ca(UO_2)_2Si_2O_7.6H_2O$; and autunite $Ca(UO_2)_2$ $(PO_4)_2.12H_2O$.

(b) Uranium reserves

It has already been emphasized that reserves are not known with a high degree of accuracy. To illustrate this, reference can be made to the official estimates of reserves for the non-communist countries at the end of the years 1958 and 1961. The 1958 figures were 820 000 t U but by 1961 the estimate had been reduced to 525 900 t, though only 82 015 t U had been recovered during the interval. At the end of 1972 revisions of estimates together with new discoveries resulted in reserve figures being increased to 866 000 t U (table 1), which is a net gain of only 46 000 t over 14 years. During the period 1959–72, however, 278 665 t U were produced.

TABLE 1. ESTIMATED WORLD RESERVES OF URANIUM (NON-COMMUNIST COUNTRIES); PRICE RANGE < \$20/kg U_3O_8†

country	reserves 10^3 t U	estimated additional resources/10^3 t U
Australia	71	78.5
Canada	185	190
France	36.6	24.3
Gabon	20	5
Niger	40	20
South Africa	202	8
U.S.A.	259	538
others	53	52
totals	866	916

From O.E.C.D. N.E.A./I.A.E.A. Report 1973.
† Relates to present-day prices.

Conversion factors commonly used in uranium and thorium resource assessment

1 tonne (t) = 0.9842 long ton (or U.K. ton)
= 1.1023 short ton
1 tonne U = 1.300 short ton U_3O_8
1 tonne U = 1.179 tonne U_3O_8
1 % U_3O_8 = 20 lb U_3O_8/short ton
= 10 kg U_3O_8/tonne

(i) Canada

The bulk of the reserves of Canada occur in peneconcordant deposits in quartz-pebble conglomerates near the base of the Huronian Supergroup in the Elliot Lake–Blind River region. The ore-bodies are elongate or tabular and vary greatly in size. On average the economic horizons are about 3 m thick but they range from 2 to 12 m and contain an average of about 0.1 % U_3O_8 in currently mineable areas. The main radioactive minerals are uraninite and brannerite, but minerals such as monazite and uranothorite are present and variations in the relative proportions of all these minerals result in changes in the uranium:thorium ratio of ore from 4:1 to 1:4 between different mines. The thorium-rich ores tend to be more refractory and therefore more costly to up-grade. Reserves are relatively easy to assess by drilling, although some are deep-seated, and are estimated to be about 150 000 t U.

The remaining 35 000 t of Canada's reserves are divided between vein-type desposits in the Beaverlodge area of Saskatchewan, British Columbia, Northwest Territories and Newfoundland. Limited reserves – probably of the order of 2000 t U – also occur in the granite pegmatites of the Bancroft area of Ontario. The most significant discoveries of uranium in Canada in recent years have been of vein and replacement orebodies in the Carlswell–Wollaston Lake area of northern Saskatchewan.

(ii) *United States of America*

The uranium province of the Colorado Plateau, the Middle Rocky Mountains and the Great Plains is the most important in the U.S.A. The deposits occur essentially in fluviatile quartzoze sandstones and mudstones and vary in size from small pockets to masses several hundred metres in lateral dimensions and up to 30 m in thickness. The larger deposits are commonly about 3 m thick and contain more than 10^6 t of ore at a grade of 0.2 % U_3O_8. The bulk of the reserves is contained in deposits less than 100 m deep and nearly half of the reserves could be mined by open-pit methods. Similar though much smaller deposits occur in the Gulf Coastal Plain of Texas. Reserves total 250 000 t U.

The main uranium minerals are uraninite and coffinite though many secondary minerals occur near the surface, particularly carnotite. Ore is readily amenable to treatment and chemical concentrate can therefore be produced relatively cheaply.

Deposits in the form of veins, stockworks and mineralized breccias occur at several locations, the major ones being in the Colorado Mountain area, the Basin and Range Province and the Northern Rockies. The reserves together amount to approximately 9000 t U.

(iii) *Republic of South Africa*

The uranium deposits of the Witwatersrand basin are the most important in the African continent. They are peneconcordant and essentially confined to quartz-pebble conglomerates somewhat similar to those of the Elliot Lake–Blind River field. They differ mineralogically, however, in that the only important uranium mineral is a low-thorium uraninite with a Th:U ratio of $\approx$ 1:33. The host sediments are about 8000 m thick in the central part of the basin and contain five main uranium-bearing horizons. The grade of the ore averages 0.025–0.03 % U_3O_8 and at a selling price of < $20/kg U_3O_8 is available only as a by-product of gold extraction. Reserves are about 114 000 t U.

Uranium also occurs in alaskitic pegmatites at Rössing near Swakopmund in southwest Africa. The main ore mineral is uraninite but a number of secondary minerals are present near the surface. Official figures of reserves and grade have not been released but they are quoted as being 77 000 t at an average grade of 0.035 % U_3O_8.

A relatively small tonnage of uranium occurs in uranothorianite, which is present as an accessary mineral in the carbonatite complex of Palabora in North East Transvaal. Reserves are 11 000 t though the grade (0.005 % U_3O_8) is such that uranium can only be recovered as a by-product.

(iv) *Australia*

In 1970 uranium reserves in Australia were estimated to be 16 700 t U contained mainly in the pyrometasomatic deposit of Mary Kathleen in Queensland and in smaller vein-type deposits in Northern Territory. Since then, however, active prospecting in favourable geological environments has resulted in a succession of discoveries in four main provinces: Rum Jungle–Alligator Rivers; Mary Kathleen–Westmoreland; Mount Painter–Radium Hill; and Kalgoorlie–Wiluna.

The deposits of the Rum Jungle–Alligator Rivers province occur in faulted and folded metasediments of Lower Proterozoic-Carpentarian age. They are of medium tonnage and relatively high grade (≈ 0.25 % U_3O_8). Reserves have not yet been established but indications are that together they are likely to contain in excess of 100 000 t U. Deposits in the Westmoreland area are essentially of pitchblende in sheared sediments and each have only a few thousand tonnes U.

In South Australia uranium occurs at Lake Frome and Yrramba in sandstone deposits similar to those of the Colorado plateau. The ore mineral is finely divided pitchblende and grades are of the order of 0.2 % U_3O_8. Reserves are not yet firmly established but are estimated to be more than 20 000 t U.

The recent discovery of a type of uranium deposit new to Australia at Yeelirrie 80 km southwest of Wiluna, Western Australia, considerably enhances Australia's role as a potential supplier of uranium. The company reports reserves of 40 000 t U at an average grade of 0.15 % U_3O_8.

(v) *France*

The main uranium deposits of France are vein-type with pitchblende the main uranium mineral. Reserves are officially given as 36 600 t U, 75 % of which are in the districts of Limousin, Forez and Vandée. The remaining 25 % occur as peneconcordant deposits of Permian age in the Hérault basin.

(vi) *Gabon*

The peneconcordant deposits of Mounana and Oklo occur in feldspathic sandstones intercalated with conglomerates but are strongly controlled by faulting. Uranium is present as pitchblende and several secondary uranium minerals. Reserves are 20 000 t U.

(vii) *Niger*

Uranium in Niger occurs as peneconcordant deposits in Carboniferous and Cretaceous sediments of the Agadès basin, bordered to the east and north by Precambrian crystalline rocks of the Aïr massif and to the northwest by the Hoggar massif. Ore grade material at Arlit averages 0.25 % U_3O_8 and the disposition of the orebodies are such that mining of much of the ore can by by open-pit methods. Reserves are estimated to be 40 000 t U.

(viii) *Other countries*

Other countries with reserves of more than 1000 t U include Argentina, Central African Republic, Greenland, Italy, Japan, Mexico, Portugal, Spain, Turkey, Yugoslavia and Zaïre. Reserves in all of these countries together total 52 700 t U.

3. AVAILABILITY OF RESERVES

Attention was first drawn to the problem of availability of reserves at the Third International Conference on the Peaceful Uses of Atomic Energy (Mabile & Gangloff 1965; Bowie 1965). The most obvious restriction to output results when uranium is mined as a by-product. However, there can also be physical difficulties due to the geometry of the ore deposit or to the environmental effect of mining low-grade ore in a developed country.

Approximately half of the reserves (75 000 t U) of the Elliot Lake–Blind River province are in the property of Denison Mines Limited. Mill capacity was initially 5500 t/day, but this is being extended to 6800 t/day. It could no doubt be increased further to 9000 t/day; but it seems doubtful if this would result in production exceeding 4000 t U/year. Canada's planned capacity is for an output of 6500 t U/year, and to extend this beyond 11 000 t U/year would mean the development of additional reserves.

TABLE 2. WORLD URANIUM PRODUCTION (TONNES U)

country	1959 (peak year)	1972
Argentina	—	26
Australia	860	—
Canada	12 243	4 003
Congo (Zaïre)	1 786	—
France	897	1 380
Gabon	—	210
Japan	—	15
Niger	—	870
Portugal	115	81
South Africa	4 966	3 076
Spain	—	60
Sweden	—	7
U.S.A.	13 398	9 900
totals	34 265	19 628

Partly based on O.E.C.D. N.E.A./I.A.E.A. Report 1973.

There are no major problems of uranium availability in the U.S.A. because of the nature of the deposits, though recovery will become more difficult as underground mining increases. But in the case of the Republic of South Africa it seems unlikely that more than 5000 t U could be produced annually as a by-product of gold from the Witwatersrand. Production from open-cut mining at Rössing will help boost South African output even though the grade is low.

It is too early yet to comment on possible production difficulties in Australia but there would appear to be no serious problems. It will be necessary to control radon build-up at properties where the average grade is high, such as at Nabarlek, by making sure that adequate ventilation facilities are installed.

Production in France could not be raised to more than 2000 t U/year unless significant new finds are made. In the Gabon, production is likely to be limited to 1200 t U/year. Output from Niger will probably be of the same order, being restricted by high transportation costs and the mining of additional ore underground. Present production is compared with the peak producing year of 1959 in table 2. Estimates of the uranium that could be available from the seven main and other potential producing countries listed above are given in table 3. This estimate equates with the medium-range forecast and would indicate the possibility of demands being satisfied to the beginning of the 1980s. It is of major importance, however, to note that the predicted demand – allowing for the introduction of fast breeder reactors – is likely to more than double from 1980 to 1990 (figure 1).

TABLE 3. ESTIMATED CAPABILITY OF URANIUM PRODUCTION BY 1980

country	annual production capacity/tonnes U
Australia	10000
Canada	11000
France	2000
Gabon	1200
Niger	1500
South Africa	7500
U.S.A.	25000
others	2000
totals	60200

Based on O.E.C.D. N.E.A./I.A.E.A. Report 1973.

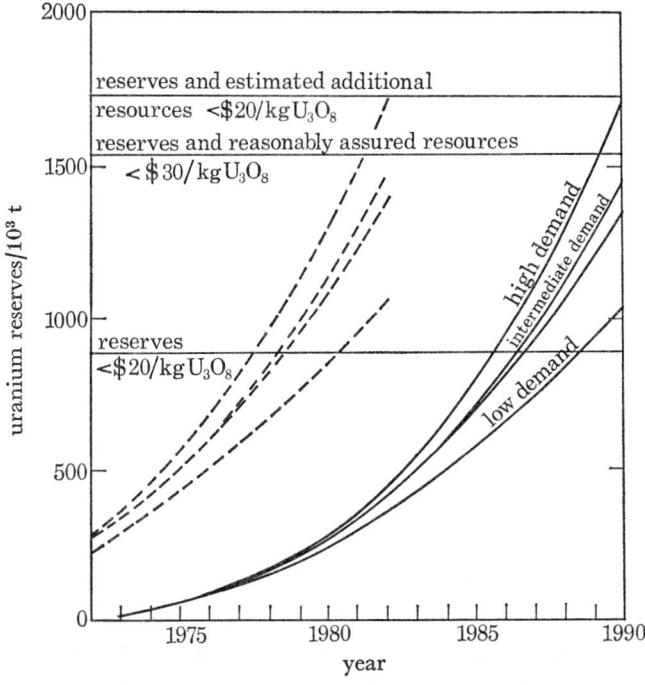

FIGURE 1. Relationship between uranium reserves, resources and estimated cumulative requirements. The dashed curves allow for an 8-year forward reserve. (Based on O.E.C.D. N.E.A./I.A.E.A. Report 1973.)

4. ESTIMATED ADDITIONAL RESOURCES AT $< \$20/\text{kg}\ U_3O_8$

The estimated additional resources possibly available but not yet known with any certainty in the seven main producing countries total 864000 t U with 52000 estimated as being present in other countries (table 1). These resources have in many instances to be delineated, upgraded to reserves and made available at the required rate. To sustain an output of 150000 t U/year, reserves would require to be approximately three times those currently estimated. The obvious question is where are they to be found?

No doubt new reserves will be discovered in known producing areas and in new uranium provinces. However, the task of doing so is not going to prove as easy as it was during the 1950s when most of the main uranium provinces presently known were recognized.

5. REASONABLY ASSURED AND ESTIMATED ADDITIONAL RESOURCES AT $20–30/kg U_3O_8

Reasonably assured and estimated additional resources in the non-communist countries total 1 312 000 t U (table 4), with by far the largest proportion in Canada, Sweden, the U.S.A. and South Africa. The low grade of these deposits makes the problems of availability more acute than in the case of conventional ore deposits. For example, the extensive occurrences in the Västergotland and Närke districts of Sweden contain about 10^6 t U_3O_8, but because of mining and treatment losses the resources are estimated at 310 000 t U – which is a realistic figure – but for environmental reasons it is not considered that output is likely to exceed 2000 t/year.

TABLE 4. ESTIMATED WORLD RESOURCES OF URANIUM (NON-COMMUNIST COUNTRIES);
PRICE RANGE $20–30/kg U_3O_8†

country	reasonably assured resources/10^3 t U	estimated additional resources/10^3 t U
Australia	29.5	29
Canada	122	219
France	20	25
Niger	10	10
South Africa	62	26
Sweden	270	40
U.S.A.	141	231
others	26	52
totals (rounded)	680	632

From O.E.C.D. N.E.A./I.A.E.A. Report 1973.
† Relates to present-day prices.

6. FUTURE OUTLOOK

Currently known reserves are adequate to supply the predicted requirements to 1980 if adequate steps are taken in time to regenerate or rebuild plant that has become obsolete during the period of hiatus in production that is now showing signs of having ended. In order that continuity of supply can be assured, however, adequate forward reserves must also be established. This lead time is normally accepted as being a minimum of eight years. For < $20/kg U_3O_8 reserves, this means that more than twice the reserves established since the early 1950s must be found in approximately half the period of time that was then available. This is such a formidable task that it is predicted that uranium could well have to be recovered from resources in the $20–30/kg U_3O_8 range during the latter part of the 1980s.

7. Thorium reserves and resources

Information available on thorium reserves and resources are less reliable than for uranium as there has been no significant requirement of thorium for nuclear reactors. High-temperature gas-cooled reactors have been constructed that operate on a thorium fuel cycle and if early promise is upheld, thorium could become an important nuclear fuel of the future. Reserves and estimated additional resources at $<$\$20/kg ThO_2 are given in table 5. Important ore minerals are monazite (Ce, La, Nd, Th) PO_4, thorite $ThSiO_4$, uranothorite (U, Th)SiO_4 and brannerite.

Table 5. World reserves and resources of thorium (non-communist countries); price range $<$\$20/kg ThO_2

country	reserves 10^3 t Th	estimated additional resources/10^3 t Th
Brazil	1.2	31.8
Canada	80	80
Arab Republic of Egypt	14.7	280
India	—	300
South Africa	20	—
U.S.A.	52	265
others†	8	—
totals (rounded)	176	957

Based on O.E.C.D. N.E.A./I.A.E.A. Report 1973.
† Mainly in Australia, Korea, Malagasy Republic, Malaysia, Malawi, Nigeria, Sierra Leone.

Production in the past has been mainly from placer and vein deposits, but in recent years the element has been recovered as a by-product of uranium from the Elliot Lake–Blind River uranium deposits. Canadian reserves and resources in the $<$\$20/kg ThO_2 price range are estimated to be 160000 t Th. Even larger reserves and resources occur as beach deposits in India, where there are 300000 t Th. Large reserves (14700 t Th) with resources estimated at 280000 t Th also occur in beach deposits in the Arab Republic of Egypt. Similar deposits occur in Brazil, where reserves and resources total 33000 t Th. South Africa has large reserves and resources in detrital monazite in Karoo sediments and in vein deposits in Cape Province – which in the mid 1950s produced over 8000 t monazite/year. In the U.S.A. thorium occurs in the minerals thorite and monazite in vein deposits in Colorado, Idaho and Montana. These contain reserves of approximately 37000 t Th and much of the estimated additional resources of 265000 t Th of the U.S.A. The remaining 15000 t Th reserves are in placer deposits in Idaho, Montana and North and South Carolina.

8. Conclusions

Presently known reserves in the price category of $<$\$20/kg U_3O_8 amount to 886000 t U. To this can be added about 100000 t U, which will almost certainly be established as reserves in Australia in the near future. There is therefore adequate uranium defined to ensure that there should be no shortages in the 1970s. By 1980 it is estimated that uranium available from known sources could reach 60000 t U/year, which is the same as the medium-range forecast figure for that year's requirement (table 6). There is likely to be little spare production capacity unless immediate steps are taken to ensure that output can be substantially increased by that time.

It is estimated that demand will be nearly double the 1980 figure by 1985 and continue to increase to around 160000 t/year by 1990. To ensure that production keeps pace with demand, and that an 8-year forward reserve is maintained, is a task that will tax the abilities of geologists, mining engineers and metallurgists alike over the coming years.

Accumulative requirements are forecast as being about 1400000 t U to 1990. and with an 8-year forward reserve this will mean the discovery of more than twice presently known reserves by 1982. How can this be done?

TABLE 6. 'BEST' ESTIMATE OF ANNUAL URANIUM REQUIREMENTS
(NON-COMMUNIST COUNTRIES)

| year | 10^3 tonnes U | | year | 10^3 tonnes U | |
	annual	cumulative		annual	cumulative
1973	17	17	1982	77	417
1974	20	37	1983	85	502
1975	25	62	1984	95	597
1976	30	92	1985	105	702
1977	35	127	1986	116	818
1978	40	167	1987	128	946
1979	45	212	1988	140	1086
1980	60	272	1989	151	1237
1981	68	340	1990	164	1401

Based on O.E.C.D. N.E.A./I.A.E.A. Report 1973.

Nearly all of the known uranium provinces of the world were discovered during (or had been recognized before) the era of intensive search for uranium between 1945 and 1958. This applies even in the case of Australia, although the reserves in that country were not considered to be as large as is now anticipated. The fall of discovery rate of new uranium deposits during the 1960s resulted from two main reasons: lack of demand and the fact that a high percentage of deposits with surface manifestations had already been discovered in readily accessible parts of the world. The present erosional surface of the Earth is fortuitous so far as uranium deposits are concerned and it is to be expected that many more deposits than have been discovered are too deeply buried to be detected by γ-ray measurements. To ensure that such deposits are discovered will mean increasing basic research relating to the geology, mineralogy and chemistry of known uranium occurrences; improving our understanding of regional and local controls of ore deposition; and developing geophysical and geochemical methods of detecting orebodies tens to hundreds of metres below surface.

In the field of extractive metallurgy, continued and intensified research is also necessary into new methods of beneficiation. For example, it has recently been shown (Davis et al. 1972) that mechanical sorting of ore, bacterial leaching, and new methods of chemical leaching can assist appreciably in reducing the costs of extracting uranium from marginal grade ore. Much more can be done on other extractive techniques including leaching of ore in situ.

Presently negotiated prices for uranium concentrate ($12–16/kg U_3O_8) are too low to ensure that known orebodies are not high-graded leaving the remainder in the ground and thus reducing reserves. Furthermore, the cost of prospecting for uranium, which is estimated to have averaged about $2/kg U_3O_8, will increase as discovery becomes more difficult. With no real short-term economic incentive to prospect, which is a situation pertaining in almost every country with the exception of Australia, the risk of a shortage of uranium in the 1980s is increased. For this reason, the sooner the market recovers from the present position of over-supply, the greater is the chance of adequate uranium being found in time to meet future demands

In the event of all precautions being taken and a shortage of uranium still appearing a probability, recourse can be made to the recovery of uranium from submarginal raw material from which well over 10^6 t U could be recovered. Should this still prove inadequate, there are further resources in the price category of \$30–60/kg U_3O_8 that could be made available from conventional type deposits of low grade in Australia, Brazil, Canada, South Africa and Spain. In addition, uranium could be obtained from the phosphate-leached zone of the Florida phosphate deposits, as well as in limited quantities from sea water and as a by-product in the manufacture of triple super-phosphate from phosphate rock.

Existing uranium stockpiles – not all officially disclosed but probably amounting to 80 000 t U – could act as a buffer against any unpredicted shortage, but this is certainly not a time for complacency so far as future supplies are concerned. The lesson of diminishing additions to ore reserves per metre drilled in the U.S.A. in 1972 (U.S.A.E.C. Report, 1973) should make it clear that new uranium provinces must be located. The aim should be to encourage an immediate and major increase in prospecting in geologically favourable areas and to develop new techniques of discovering hidden sources of present-day ore grades. If adequate steps are taken, and in time, a sufficiency of uranium should be available to supply the needs of the 1980s and well beyond. The possible utilization of thorium and the introduction of breeder reactors should prevent any shortage developing this century, but only if adequate foresight is exercised throughout the many branches of the uranium industry.

This paper is published by permission of the Director, Institute of Geological Sciences, London.

References (Bowie)

Bowie, S. H. U. 1965 In *Proc. 3rd Int. Conf. peaceful Uses atom. Energy, Geneva* 1964 **12**, 28–31. New York: U.N.
Davis, M., Bowie, S. H. U., Ostle, D., Loosemore, W. R., Smith, S. E. & White, P. A. 1972 In *Proc. 4th Int. Conf. peaceful Uses atom. Energy*, 1971, pp. 59–72. New York: U.N. Vienna: I.A.E.A.
Lewis, W. B. 1972 *Proc. R. Soc. Edin.* A, **70**, 219–223.
Mabile, J. & Gangloff, A. 1965 In *Proc. 3rd Int. Conf. peaceful Uses atom. Energy, Geneva* 1964, **12**, 3–8. New York: U.N.
Miller, D. S. & Kulp, J. L. 1963 *Bull. geol. Soc. Am.* **74**, 609–630.
O.E.C.D. N.E.A./I.A.E.A. 1973 *Uranium resources production and demand*. Paris: O.E.C.D.
Stieff, L. R., Stern, T. W. & Milkey, R. G. 1953 *Circ. U.S. geol. Surv.* **271**, 19 p.
U.S.A.E.C. 1973 1972 *Atomic Energy Programmes Operating and Developmental Functions*. Washington.

Discussion

Dr W. C. Marshall (*A.E.R.E. Harwell, Didcot, Berkshire*)

During the discussion, Dr Bowie was asked the question: 'What is the cost of producing uranium from sea water?' Dr Bowie said that it was about \$60/kg U but that someone from Harwell should answer that question. Dr Marshall confirmed that the work by Norman Keen and his colleagues at Harwell, up-dated to present-day values of money, suggests that uranium could be obtained from sea water at the figure Dr Bowie gave.

Phil. Trans. R. Soc. Lond. A. **276**, 507–526 (1974)

Printed in Great Britain

Geothermal power

By T. Leardini

Ente Nazionale per l'Energia Elettrica, Rome, Italy

This paper deals mainly with the development of utilization of endogenous fluids in the world and particularly in Italy, together with a forecast of the potential increase in geothermal production.

At the end of 1972 the installed capacity of the world's geothermoelectric plants was approximately 1000 MW, of which 390.6 MW are installed in Italy. In the same year the electric energy generated by Italian power stations was 2582.4 GW h. In some countries, geothermoelectric energy costs ranged from 1.4 to 2.5 U.S. mills/MJ (5 to 9 mills/kW h) as compared with 1.66–3.9 U.S. mills/MJ (6–14 mills/kW h) for alternative sources.

The total geothermal capacity in the world is expected to double and perhaps to triple in the 1980s, as new installations are being constructed or planned in several thermal areas. The utilization of geothermal fluids for evaporating low-boiling liquids (freon, isobutane, etc.) or for driving a gravimetric loop, offers attractive possibilities of using thermal waters also for the generation of electricity.

In many countries, low enthalpy fluids are used directly for other purposes (domestic and greenhouse heating, refrigeration and air-conditioning, production of fresh water, drying seaweeds and diatomites, etc.).

The cost of geothermal heat thus employed is 0.7–1.2 $/GJ as compared with about 2.6 $/GJ if fossil fuels are used. Due to this attractive cost, in the next few years there should be a remarkable development in this type of utilization of low enthalpy fluids.

Introduction

As is known, the terrestrial heat flow on the continents has an average value of about 6.3 μJ cm^{-2} s^{-1}. It is a tremendous amount of energy. If the total terrestrial heat flow could be transformed to electricity by means of a process having a 10% average efficiency (Noguchi 1970) the resulting energy would cover the forecast electricity demand up to the year 2000.

In some areas, particularly in those where magmatic bodies are at relatively shallow depth, the heat flow may reach a value which is 10–20 times its average value (Burgassi *et al.* 1970; Hayakawa 1970; Dawson & Dickinson 1970).

Part of the enormous quantity of heat carried by the magma is dispersed by volcanic eruption. Another part heats the fluids which are contained within the country rock and the overlying geological formations.

After heating, these fluids may reach the surface and give rise to hot springs, geysers and fumaroles. But if the ascending path is blocked by a cover of low-permeability rocks, recirculation of hot fluids takes place at a certain depth and a large convection system is established below the cover. The heat loss of the system to the surface is in this case rather small because heat diffusion occurs by conduction only. On the other hand, if the cover is cut by open fractures, or drilled by a well, the circulation of fluids in the deep layers will be accelerated, thermal leaks at the surface may occur and heat conveyance to the surface is mostly achieved by convection.

The heat flow carried to the surface by the fluids of exploited geothermal areas is greater than the heat flow dispersed by conduction and can attain values about 200–1700 times the average terrestrial conductive heat flow. Values of this magnitude can be considered sufficiently valuable for industrial exploitation, either for transformation into electric energy by means of conventional equipment or for heat-requiring processes. In the boraciferous region, for instance, the heat of the fluids extracted from the wells amount to 2.4 GJ/s. The average heat flow in this region (170 km²) is therefore 1410 μJ cm⁻² s⁻¹.

In the Larderello–Valle del Secolo area proper (10 km²) the heat extracted from the wells amounts to 1.06 GW, with an average heat flow rate of 10 600 μW cm⁻². Of course, although enormous, the amount of energy in a geothermal reservoir is finite. This amount can be evaluated with reasonable approximation, but the evaluation itself depends on the assumptions made regarding base temperature, porosity, pressure, etc., existing in the reservoir. This means that at the beginning of a geothermal project, in spite of the progress achieved in the pre-liminary evaluation of the reservoir capacity, some uncertainties remain on the maximum value of exploitable energy.

One could therefore conclude that the potential or recently discovered resources can only be assessed approximately. Therefore, an investment in a geothermal project always incurs a considerable initial risk.

Once a natural steam field has been located and explored, its energy can be exploited with acceptable capital costs and reasonable expense. In Italy, for instance, the average well length is 610 m, based on 670 drillings carried out up to now. The economic results of geothermal energy exploitation have been satisfactory and competitive with other sources (Facca & Ten Dam 1963; Leardini 1970).

The possibility of transporting geothermal energy by means of steam ducts is very limited, under the pressure and temperature conditions existing in the geothermal field. As a conse-quence, the heat flow carried by the fluid must be utilized directly where the wells are located. If the heat is to be utilized for industrial processes or agricultural purposes, the manufacturing plants or the greenhouse to be heated must be located near the endogenous steam-extraction area. Plants having high process heat requirements should be considered where steam is avail-able. Endogenous steam, in fact, cannot be stored in the reservoir for a long period, and cannot be transported at surface for long distance. The exploitation of geothermal steam via conversion into electricity, on the contrary, offers many advantages if it is done on the spot.

Endogenous fluids can be exploited safely and economically only when the extraction rate is continuous. The incidence of cost of the steam in geothermal electric production is low compared to the fuel cost in the other thermoelectric plants (liquid, solid or gaseous fuels, nuclear fuels). Therefore, the electric output of a geothermal plant should be constant over the year and it should be incorporated into the base load that the generating plant system is called to produce.

Up to now geothermal resources have been exploited to generate electricity in relatively limited areas at Larderello and Mount Amiata (Italy), Wairakei and Kawerau (New Zealand), The Geysers (U.S.A.), Cerro Prieto and Pathé (Mexico), Otake and Matsukawa (Japan), Pauzhetka and Paratunka (U.S.S.R.), and Namafjall (Iceland). In El Salvador, near Ahua-chapan, engineering and design work for a 30 MW power-station is under way.

Sources of endogenous hot fluids that are or could be used in the primary form (heating, and other industrial uses) are more frequent, and may be found in larger areas. The utilization of such fluids is of interest not only to the above-mentioned countries but also to many others: Algeria, El Salvador, Chile, France, Greece, Hungary, Turkey, the Philippines, Taiwan, etc. Moreover, many other countries are now starting, or have already started studies and research (see figure 1) on the identification, harnessing and utilization of endogenous fluids.

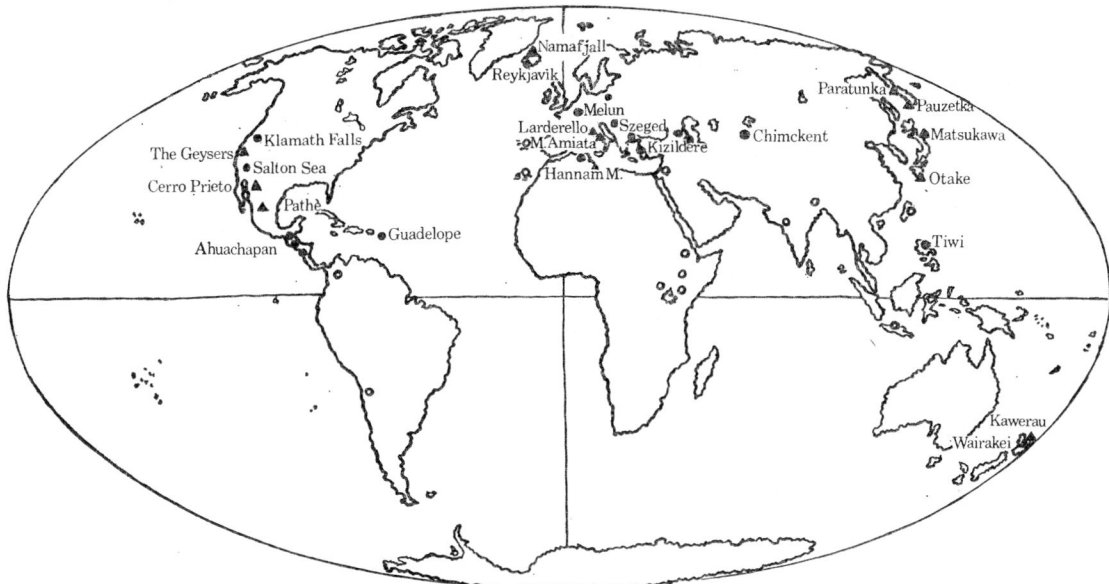

FIGURE 1. State of world geothermal research and exploitation. ▲, Countries in which geothermal power plants are in operation; ●, countries in which geothermal fluids are used for heating, agricultural and industrial purposes; ○, countries in which research is under way.

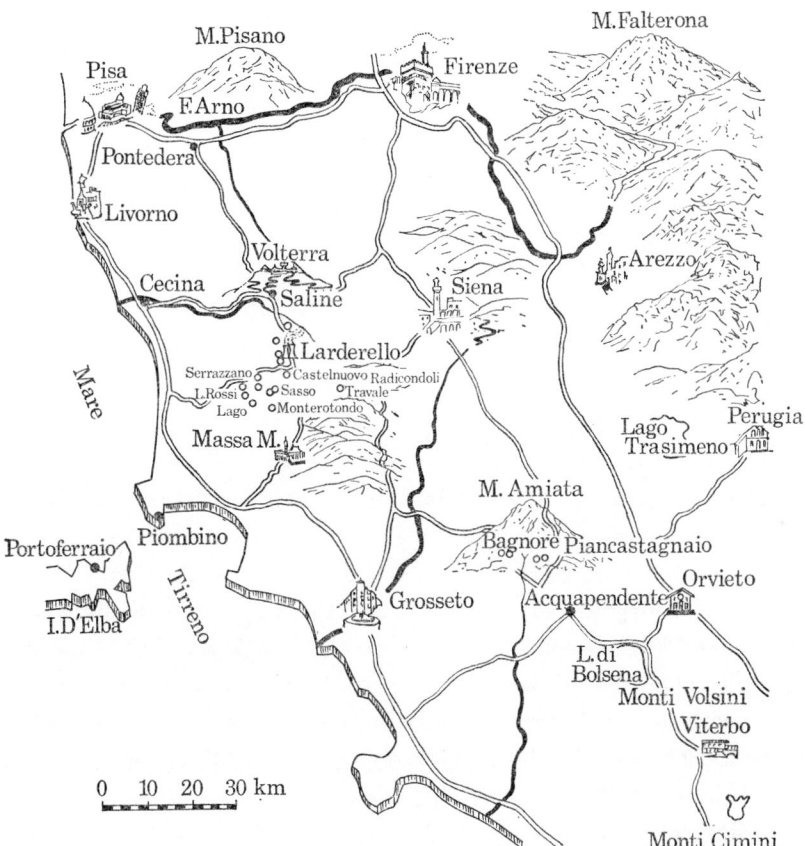

FIGURE 2. Geothermal areas in Italy. ○, Geothermal power station.

Geothermal power in Italy

Tuscany, one of the regions of central Italy, is well known all over the world for a particular industrial exploitation. In the Larderello area, about 100 km to the south of Florence (see figure 2) man succeeded for the first time in utilizing geothermal energy in industrial activity. At first (1827) natural thermal energy was used in the boric acid industry, later (1904) it was utilized for both chemical and electric production; since 1964 only electricity has been produced.

The most striking thermal manifestations in the Larderello area consisted initially of fairly strong and noisy jets of steam, the so-called 'soffioni', escaping from rock fissures and of bubbling gas in small natural pools ('lagoni') where rainwater and condensed steam had collected. The extraction of boric acid from the pools at first encountered great technological and practical difficulties, due to the lack of previous experience. The boric-bearing pool water was transferred to sedimentation tanks, fed into iron boilers and then suitably concentrated, using wood from the neighbouring forest as fuel. The solid product was obtained by means of subsequent crystallization. From 1827 onwards, natural thermal energy obtained from shallow wells was used in place of wood, which was becoming increasingly short.

The consequent outstanding increase in production of boron and ammonium compounds, the fast growth of trade and gradual growth in chemical plant capacity led, between the end of 1800 and the early decades of the present century, to the development of a flourishing chemical industry of international renown (Mazzoni 1954).

Steam exploration by means of drilling was initially concentrated in areas with natural manifestations. It was subsequently extended to a very large area in the Larderello region and was carried out with continually improved drilling equipment.

The generation of electric power with endogenous fluids was first obtained in 1904, by Prince Piero Ginori Conti: it was accomplished by admitting into a reciprocating engine the steam separated from the accompanying water. The engine exhausted to the atmosphere and was coupled to a d.c. generator which produced electric lighting for the town of Larderello.

In 1913 a 250 kW turbo-alternator replaced the reciprocating engine: geothermoelectric production on an industrial scale dates from this year. Studies carried out on the direct utilization of natural steam with condensing turbines showed that at the time the consumption of energy required for cooling-water circulation and gas extraction from the steam would have been too expensive. On the other hand, in order to extract boric acid from the steam, it was necessary to condense the natural steam in heat exchangers. These allowed the recovery of the fluid thermal energy by generating secondary pure steam for the turbines. In this manner corrosion was also eliminated and the handicap met in driving engines by direct steam was overcome.

The troubles arising from the heat exchangers and the resulting losses in energy led to experiments with impulse and back-pressure turbines directly fed by natural steam. This proved that the endogenous fluids could be utilized with direct inlet turbines exhausting to the atmosphere (cycle 1) as well as with the condensing turbines driven by pure steam produced by the aforesaid exchangers (cycle 2). The direct-inlet turbines exhausting to the atmosphere were quicker to install owing to their simple design, and cheaper to operate, as compared to the condensing units driven by indirect steam. On the other hand, the heat rate (20 kg of fluid at 500 kPa 185 °C, 5% gas content by mass, per saleable kW h net) was considerably higher than in the condensing units fed by indirect steam (14 kg of endogenous fluid per saleable kW h net).

The amount of boric acid carried by the steam gradually decreased; at the same time its economic values also decreased when rich deposits of colemanite (Turkey), rasorite and kernite (U.S.A.) were discovered and exploited. Since then, natural steam was increasingly used for power production. The growing interest in the electrical aspects of geothermal utilization led to improved drilling equipment (Giovannoni 1970) and prospecting of wider areas.

Italian equipment manufacturers began developing condensing steam turbines designed to utilize natural steam, despite its high content of incondensable gases and its carry-over of boric acid and other chemicals.

TABLE 1. GEOTHERMAL-ELECTRIC DEVELOPMENT IN ITALY

31 December ...	1920	1930	1940	1950	1960	1970	1972
number of geothermal power stations ...	2	4	5	6	11	17	17
cycle 1 power plants							
no. of turbo-alternators	—	4	5	5	5	10	10
installed capacity/MW	—	5.20	8.70	16.5	16.5	42.9	42.9
cycle 2 power plants							
no. of turbo-alternators	3	4	9	12	8	—	—
installed capacity/MW	9.25	9.25	64.25	123	79	—	—
cycle 3 power plants							
no. of turbo-alternators	—	—	—	3	14	26	26
installed capacity/MW	—	—	—	72	189.7	347.7	347.7
total no. of turbo-alternator units	3	8	14	20	27	36	36
total installed capacity/MW	9.25	14.45	72.95	211.5	285.2	390.6	390.6
average electric power (gross)/MW	0.82	6.53	60.97	145.91	239.51	311.04	294
plant capacity factor (%)	8.8	45.2	83.6	69	84	79.6	75.3
average electric power (net)/MW	0.78	5.92	55.71	134.21	217.24	287.38	269
auxiliary services and transformation loss (%)	4.9	9.3	8.6	8	9.8	7.6	8.5
fluid on stream:							
to power plants/t h^{-1}	50	150	1000	2150	2760	3400	3190
other uses/t h^{-1}	250	350	1000	70	80	100	100
average fluid rate/kg kW^{-1} h^{-1} net	64.18	25.34	17.95	16.02	12.71	11.38	11.86
cumulative net energy from 1904							
GW h	69.05	374.76	1991.80	8434.87	25515.25	49025.51	53828.43
10^{15} J	0.25	1.35	7.17	30.3	91.8	176.5	193.8
cumulative drilling from 1904/km	5	13.18	38.07	68.77	205.73	376.42	401.90
drilling per saleable GW h net/m	72.41	35.17	19.11	8.15	8.06	7.67	7.40
no. of completed wells	70	106	203	272	457	642	665
no. of wells on stream	50	60	100	123	171	199	204
average output per productive well/t h^{-1}	6	10	20	18.05	16.61	18.23	16.13
length of pipelines network/km	2	5	20	30	71.8	105	113
explored areas/km²	0.5	1	3	5	50	215	215

Mixture condensers, equipped with powerful centrifugal gas extractors (Dal Secco 1970) led to a solution of problems arising from incondensable gases, so that the direct condensation system turbine (cycle 3) rapidly took over the task of energy production. In this way it was possible to obtain an important reduction in the specific fluid consumption (an average 10 kg of fluid at 500 kPa, 185 °C, 5 % gas-content by mass is required to produce 1 saleable kW h net). The principal stages of geothermoelectric activity in Italy are summarized in table 1 and in figure 3, from which the following considerations emerge:

(1) Up to 1930 geothermal development was not achieved in a systematic way. The reasons for this were: the scarcity of geological, geophysical and geochemical information; an exploration philosophy based only on empirical criteria; the technical difficulties involved in drilling

large boreholes in heterogenous rocks having a very high temperature; the poor resistance to corrosive substances of the materials utilized in the construction of power plants, etc. As regards technology, great difficulties were found at the beginning (1930) in making suitable turbine blades, pump rotors, heat exchanger tubes, etc. These difficulties accounted for both the low production of steam and electric energy. In the first 30 years of exploitation, in fact, the plants of the whole area produced only 1.35×10^{15} J (375×10^6 kW h). The overall length drilled was 13180 m.

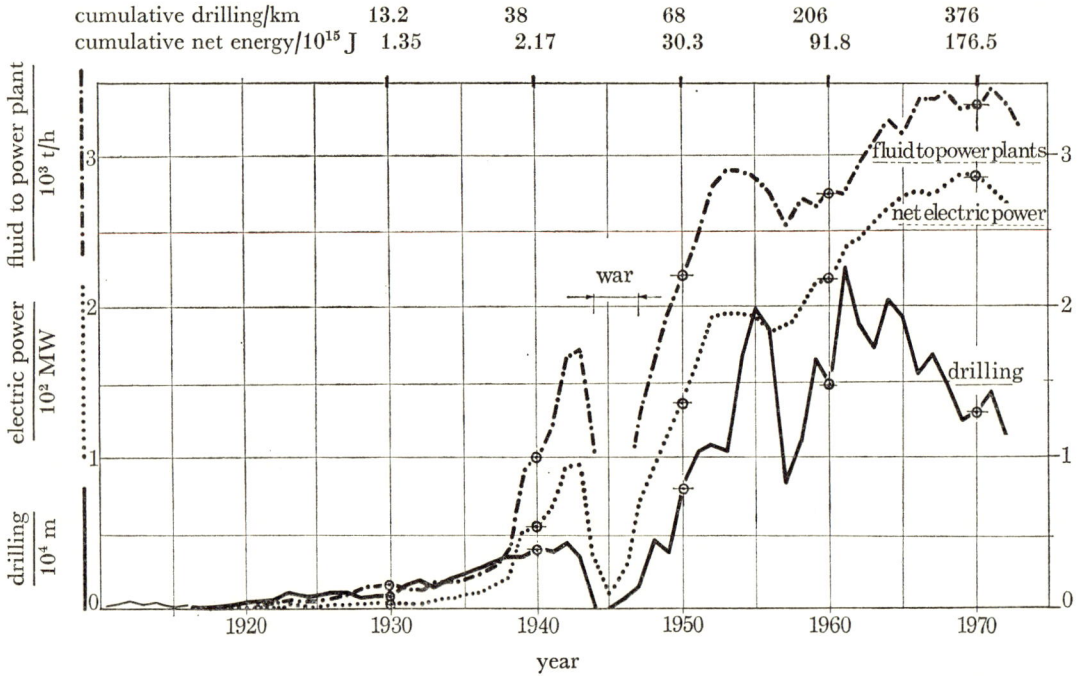

FIGURE 3. Geothermal electric development in Italy.

(2) The wells drilled in the subsequent 20 years (1930–1950) achieved a production of approximately 2200 t/h of natural superheated steam; at the beginning of their life some wells delivered up to 200 t/h of superheated steam, due to this remarkable increase in steam production. In 1938 it was possible to build and put into operation the first large geothermal power plant (Larderello 2).

(3) The decade from 1951 to 1960 is very important in the history of geothermal electric exploitation in Italy. In fact: (a) the first power plant (Larderello 3) equipped with condensing turbines directly fed by primary steam (natural steam), came into operation and a total yearly production of 7.2×10^{15} J (2×10^9 kW h) was attained; (b) the construction of the power plants in Serrazzano (28.5 MW capacity) and Lago (19 MW initial capacity), operating with primary steam and condensing units, allowed the lowering of the average specific steam consumption from 16 to 12.7 kg/kW h and increase of the yearly plant capacity factor; (c) the discovery of new steam fields in Travale and on Mount Amiata (20 and 80 km respectively to the east and southeast of Larderello) encouraged the hope that other steam fields could be discovered outside the main Larderello area; (d) the power plant Larderello 3 was built in the years 1952–4, and in 1955 it produced over 3.25×10^{15} J (900×10^6 kW h), with a plant capacity factor of 90.4%. On the other hand, in the Travale area, despite the initial high

flow from the wells, the power plant (which was installed in 1951, with a capacity of 3.5 MW) could never reach full load and had to be stopped after a few years for shortage of steam. These two cases (Larderello 3 and Travale) illustrate that success and failure are always possible in geothermal ventures.

(4) In the years 1955–7 an unforseen event, however, stopped the increase in production of both steam and electric energy. Research and experiments proved that too short a distance between boreholes had previously been adopted, so that there was a mutual influence between wells. At the end of 1960, for example, 162 wells had been drilled over 10 km² in the Larderello area proper, from which 126500 kW were supplied.

TABLE 2. INSTALLED CAPACITY, FEEDING AND OPERATION CHARACTERISTICS OF ITALIAN GEOTHERMAL POWER PLANTS, 1972

power plant	total installed capacity MW	fluid at intake			output t/h	net capacity MW	specific consumption kg/kW h
		pressure kPa	temperature °C	gas content in mass (%)			
Boraciferous region							
1. Larderello 2	69	400	196	7	422	47	11.2
		300	166	6	106		
2. Larderello 3	120	420	195	7	690	68.5	10.1
3. Gabbro	15	740	220	10	115	12.2	9.4
4. Castelnuovo V. C.	50	460	185	13	180	23.6	14.4
		200	165	4	130		
		100	140	3	30		
5. Serrazzano	32	500	194	3.2	293	29.8	9.8
6. Lago 2	33.5	500	172	2.2	250	29.6	11.2
		200	143	1.6	80		
7. Sasso 2	15.7	500	180	3	164	16.4	10
8. Monterotondo	12.5	460	170	2	120	13.1	9.2
total condensation power stations	347.7	—	—	—	2580	240.2	10.7
9. S. Ippolito-Vallonsordo†	0.9	250	188	20	10	0.1	—
10. Lagoni Rossi 1	3.5	500	190	3	65	2.5	26
11. Lagoni Rossi 2	3	400	203	3.9	65	2.7	24.1
12. Sasso 1	7	500	185	3	65	3.2	20.3
13. Capriola	3	460	170	3.8	60	2.8	21.4
total exhausting-to-atmosphere power station	17.4	—	—	—	265	11.3	23.4
total power stations in the Boraciferous region	365.1	—	—	—	2845	251.3	11.3
Mt Amiata region							
14. Bagnore 1	3.5	300	136	7	46	1.3	35.4
15. Bagnore 2	3.5	330	142	7	54	2	27
16. Senna‡	3.5	—	—	—	—	—	—
17. Piancastagnaio	15	900	186	20	245	14.3	17.1
total Mt Amiata region power stations	25.5	—	—	—	345	17.6	19.6
grand total	390.6	—	—	—	3190	269.1	11.9

† Started up 21 December 1972.
‡ Shut down 26 November 1971 for overhaul.

(5) Since 1961 research on and exploitation of geothermal energy in Italy have been directed on a more advanced basis and clear-cut criteria of investigation and utilization (Chierici 1961; Cataldi, Ceron, Di Mario & Leardini 1970; Corti, Di Mario & Mondolfi 1970). On the basis of the preceding experience in geothermal utilization in Italy, a better knowledge of the basic phenomena and new survey methods have been resorted to for field exploration. The results of electric exploitation in Italy for the period 1960–72 may be summarized as follows: (a) The overall electric energy fed to the national electric grid was 10^{17} J $(28 \times 10^9$ kW h); (b) The average production was **3300** t/h of superheated steam; (c) The average drilling was approximately 16 300 m/year; (d) The length of new steam pipelines (with diameter ranging between 250 and 810 mm) was 41 km.

In the period considered, the turbines operated with secondary steam have been abandoned. In general, they have been replaced by condensing turbines fed by natural steam. In some areas, however, where the high content of incondensable gases (20–80 % by mass) does not economically justify a condensing cycle (Zancani 1959) a few back-pressure exhausting-to-atmosphere turbines still operate.

A list of power plants in operation in 1972 and their features is given in table 2.

GEOTHERMAL ENERGY IN OTHER COUNTRIES

Until 1958 Italy was the only country where natural steam was used on an industrial scale for power generation. Since then, other countries have started to exploit geothermal fluids (Facca 1970). At the end of 1972 the installed capacity of the world geothermoelectric plants was as follows:

Italy		
Larderello	365 100 kW	
Monte Amiata	25 500 kW	
total		390 600 kW
U.S.A.		
The Geysers		298 000 kW
New Zealand		
Wairakei	192 000 kW	
Kawerau	10 000 kW	
total		202 000 kW
Mexico		
Pathé	3 500 kW	
Cerro Prieto	75 000 kW	
total		78 500 kW
Japan		
Otake	13 000 kW	
Matsukawa	20 000 kW	
total		33 000 kW
U.S.S.R.		
Pauzhetka	5 000 kW	
Paratunka	750 kW	
total		5 750 kW

Iceland
Namafjall   3 000 kW
    grand total   1 010 850 kW

The countries using hot water and/or steam for heating and industrial uses are:

Algeria. Hot springs at a temperature of 90 °C are utilized at Hanmam Mescontine, southwest Annaba: deep drilling is under way in Sidi Zid.

Chile. Natural steam has been found in the El Tatio area and production wells have been programmed. The Chile project provides for threefold utilization of geothermal resources: power production, mineral extraction and desalination.

China. Geothermal energy in the form of natural steam is used for electricity generation in a number of areas. In some urban areas, natural hot water has been piped for heating purposes and for industries requiring a large supply of hot water (dye factories, paper mills, chemical works). In rural areas warm and hot springs have been channelled for use in nurseries and greenhouse, irrigating seedlings, saplings and vegetables. They are also utilized in poultry hatcheries and in certain fermentation processes (Takung Pao 1972).

France. In Guadaloupe, one geothermal field has been located. It produces a mixture of steam and water at a present flow rate of about 140 t/h at 160 °C. In the region around Paris a project is under way to provide domestic heating in Melun. The geothermal fluids are produced in the nearby region.

Hungary. Eighty hot water wells have been in operation since 1969. The total flow rate is 6800 t/h. The water temperature is 85–95 °C. The heat is utilized for heating 1200 homes in Szeged (southern Hungary) and greenhouses covering an area of 400000 m² (Boldizsar 1970).

Iceland. The most important utilization of geothermal fluids is that of space heating. Public administration buildings, as well as private homes, in Reykjavik and suburbs are centrally heated by endogenous fluids. More than 40 % of Iceland's population live in homes heated by geothermal fluids.

Recent boreholes have increased potential wellhead energy (above 40 °C) for house heating to 1130 GJ/h (National Energy Authority, Reykjavik, 1973).

Greenhouses covering 110000 m² are heated by the above fluid, which yields 150 GJ/h. The annual capacity factor is 35–40 %. A diatomite plant for the production of dry diatomaceous earth was built in 1966–7 near Lake Mayvatn: the plant produces 42000 t/year of diatomite and utilizes 50 t/h of fluids at a pressure of 700–800 kPa (Lindal 1970).

Japan. The fluids are also utilized for agricultural purposes (8000 m² greenhouses, poultry-farms, fish, and alligator hatcheries). The most impressive utilization of hot springs is that for recreational and therapeutic purposes: about 150 million people visit hot springs annually and over $870M are spent per year to this purpose (Komagata *et al.* 1970).

Mexico. The fresh water needed for construction and operation of the Cerro Prieto plant is produced by a pilot distillation plant fed by hot fluids. It is likely that more distillation units will be installed to cope with the needs of the surrounding semi-desert area.

New Zealand. In the Kawerau area, 75 km northeast of Wairakei, a mixture of steam and hot water is harnessed through a heat exchanger which produces secondary steam for the processing line of a nearby paper mill.

Philippines. Production wells are being drilled in the Tiwi (Luzon area): a small 2.5 kW pilot generator has provisionally been installed. At the same time a new geothermal field has been discovered on the island of Leyte.

Turkey. In 1968 the Kizildere geothermal field was discovered: one well has a potential of 30 t/h of dry steam and 250 t/h of hot water at a well-head pressure of 400–500 kPa. The Turkish Government is planning to build a geothermal plant as soon as possible (Ten Dam & Erontöz 1970).

U.S.A. The Salton Sea area (California) delivers very hot (340 °C) brine, carrying dissolved salts of many important elements (copper, uranium, silver, etc.) A pilot plant has therefore been installed to obtain the valuable elements from the salt brine. Heat is also used in different areas for air-conditioning and other purposes. At Klamath Falls (Oregon), for instance, 400 buildings are air-conditioned by hot water (100 °C) yielded by about 350 wells, drilled near the city.

U.S.S.R. Hot water (80–130 °C) sources are easily found in certain regions at very shallow depths. In the very cold region of Kamehatka, near Paratunka, a pilot power plant was installed in 1967 utilizing this kind of low-enthalpy source. The fluid operating the turbine is Freon 12, and is evaporated in heat exchangers and liquified in a condenser (Moskvicheva 1970). In Makhach-Kala, Grozni, Chimckent and Kamchatka regions some hothouse centres with an area of more than 150000 m² are being built.

TECHNICAL AND ECONOMICAL CONSIDERATIONS

The production of geothermoelectric energy largely depends on the size of the natural resources, which vary greatly from one country to another. It is therefore difficult to find common criteria valid for an assessment of the technical and economical advantage of building power plants.

The first difficulty lies in assessing the amount of geothermal fluid available for exploitation and its characteristics, e.g. its energy content. In some fields (Larderello, The Geysers and Matsukawa) the fluid is superheated steam, carrying incondensables in an amount which may differ greatly from one site to another (6 % by mass at Larderello, 0.6 % at The Geysers, 1.7 % at Matsukawa). In other fields the fluid is dry saturated steam, or a mixture of steam and water (Wairakei, Cerro Prieto and many others) with a low incondensable gas content (0.5–0.7 % by mass). Some other fields (Paratunka and Reykjavick, for example) deliver hot water at a temperature near or slightly above boiling point (80–120 °C). Finally, the Salton Sea area delivers very hot (340 °C) brine, carrying a large quantity of dissolved salts.

Another figure which is difficult to assess is the specific density of electric energy which can be produced in a geothermal area. In effect, the possibility of yielding geothermal fluids from the reservoir differs greatly from one region to another and frequently also from one area to another in the same region. For instance, taking into account the specific energy density of the Larderello region, in the main production area of Larderello–Valle del Secolo, about 10 km², a peak value of 12.5 MW/km² is attained; but the average specific energy density in the whole region (about 170 km²) is 1.6 MW/km². At The Geysers, the present steam supply (produced from an area of about 85 km²) seems to be sufficient to operate a capacity of 710 MW (McMillan 1970), which brings the average specific density to approximately 8.4 MW/km². At Cerro Prieto, the drilled area (4 km² (Guiza 1973)) can at present supply fluid to feed 75 MW, i.e. an average specific density of 18.7 MW/km².

As regards the cost of endogenous fluids, the recent data available are contained in the report of the Thermal Power Co., who deliver endogenous steam to P.G. & E. for generation of electric energy in The Geysers power plants. This report states: 'The price of steam...is up 10.41 % over 1972 to 2.65 mills. Additionally, we receive 0.5 mill per kW h for effluent disposal which brings the total payment to 3.15 mills per kilowatt hour' (Annual Report Thermal Power, 1972). This price (1 mill = 0.1 cent) includes the cost of research, survey and exploitation, steam transport to the power plant, the disposal or reinjection of cooling water, taxes, amortization, general costs and benefits. The steam is delivered to the power plant at about 500 kPa, 175 °C, and contains 0.6 % by mass of incondensables.

Ragnars, Saemundsson & Benediktsson (1970) evaluate the cost of steam delivered to the power plant Namafjall at 20 U.S. cents per tonne. When revising these costs to take into account devaluation and auxiliary services, this cost increases to 24.8 cents per tonne of saturated steam at 1 MPa, 181 °C, 0.2 % by mass of incondensables. On the basis of the average specific steam consumption (10 kg/kW h) achieved at The Geysers plant, the cost per kW h at Namafjall would come to 2.45 mills (0.66 mill/MJ), which is in very close agreement with the 2.65 mills (0.74 mill/MJ) of the selling price of Thermal Power Co. in the United States.

A recent report by Cataldi, Di Mario, & Leardini (1973) gives an evaluation of the experience in Italy during some 70 years of exploitation. When approaching a new geothermal project, some figures are foreseeable, based on this experience.

Preliminary investigations and drilling of four or five wells may cost U.S. $0.2–1M respectively. If preliminary drilling confirms favourable conditions at depth, the quantity of fluid which may be expected from the exploration wells is of the order of 80 t/h, once the flow rate has stabilized. If the cost of the steam-ducts is to be added (which can be assumed at U.S. $0.1M), the total expenditure is U.S. $1.3M.

By applying to this cost the same amortization percentage indicated by McMillan (1970) for The Geysers power plants (13.5 %), the resulting total cost for a tonne of superheated steam (500 kPa, 185 °C, 6 % by mass of incondensables) is 25 U.S. cents. This brings the cost to 0.7 mill/MJ (2.5 mills/kW h) when condensing equipment for the electric plants is adopted.

The cost of a power station, however, can be determined more accurately. The cost per unit power in geothermal power plants is approximately the same as for a fossil fuel power plant. This cost, however, may differ considerably from one area to another (even in the same country) owing to the following reasons:

(1) First of all, cost is influenced by the steam conditions. Plants utilizing water–steam mixtures have additional expense in drying cyclones, draining devices, multiple-inlet turbines, etc. (as in the dase of Wairakei), which involve higher costs than those of plants utilizing superheated steam (Larderello, The Geysers, etc.).

(2) A high gas content in the geothermal fluid requires bigger gas extractors, whose capital cost may amount to 20 % of the total.

(3) A sizeable content of chlorides and other corrosive chemicals requires the use of expensive corrosion resisting metals for steam ducts, turbine blades, pumps, extractors, etc.

(4) Another cost item to be assessed for each site is that of the cooling towers, needed when the power plant is installed far from natural lakes or rivers, as often happens.

(5) The local standard capacity may affect the costs. In countries necessitating purchase from abroad, shipping fees, customs, etc., also affect the costs and may result in a large increase in installation costs.

(6) The prevailing social and economic conditions also influence the costs, because of the different cost of money, amortization periods, etc., and the difference may be quite important.

(7) As regards the plant construction cost, the literature gives plenty of information and figures. By updating data reported in the literature (U.N. 1970), the actual cost per kW of power plant can be given approximately as follows:

Condensing-turbine power plants

Larderello

net capacity 2×26 MW: 170 U.S. \$/kW;

net capacity 1×15 MW: 226 U.S. \$/kW;

The Geysers

net capacity 2×55 MW: 125 U.S. \$/kW;

net capacity 2×28 MW: 135 U.S. \$/kW;

Back-pressure power plants

Larderello

net capacity 1×15 MW: 95 U.S. \$/kW;

net capacity 1×4 MW: 105 U.S. \$/kW;

Namafjall

net capacity 1×3 MW: 60 U.S. \$/kW.

The higher cost per kW of the Italian condensing power plants is mainly due to the fact that very large turbo-compressors are needed for the extraction of incondensable gases from the condensers. These large turbo-compressors are necessary because the incondensable gas content at Larderello is up to 10 times higher than that at The Geysers field. The higher specific cost is also partly explained by the increase in labour cost at Larderello for the reasons which will be explained later on.

The first back-pressure power plant at Namafjall shows a rather low installation cost. This, however, is attributable to the fact that it is equipped with a very simple turbine, which was recovered from an earlier plant.

As regards the operating costs, it is not easy to give figures of general validity. These values change greatly from one country to another, because of the differences in labour legislation, in the basic wages and salaries and in the type of social welfare schemes. To take one example, social costs are very high in the Larderello boraciferous region, because of its remote location. Housing, schools, hospitals, sports grounds, entertainment must all be provided by the Electric Utility, exploiting the endogenous steam in Larderello and some other power plants (Serrazzano, Lago, Sasso Pisano, etc.). Automation and remote control have been applied in almost all the Italian exhausting-to-atmosphere power plants many years ago (Di Mario & Tagliabue 1958), but they have been extended to some condensing units only in 1969. This is one of the reasons for the higher operating costs of Italian stations.

As regards the high-capacity sets (2×26 MW), the operating cost is on the average 0.55 mill/MJ (2 mills/kW h) net, and it is somewhat more for smaller units. The exhausting-to-atmosphere power plants, which are technologically less sophisticated, have an operating cost of about 0.45 mill/MJ (1.6 mills/kW h) net. Finally, for the remote-controlled sets, the operating cost is further reduced to 0.166 and 0.22 mill/MJ (0.6 and 0.8 mill/kW h) net for exhausting-to-atmosphere and condensing units, respectively.

McMillan (1970) estimates that the operating cost of the highly automated units 5 and 6 of The Geysers plant is some 0.125 mill/MJ (0.45 mill/kW h) net. This figure, when updated, is of the order of 0.147 mill/MJ (0.53 mill/kW h) net.

Hayashida (1970) reports, for the Otake power plant (13 MW capacity), a cost of 0.166 mill/MJ (0.6 mill/kW h) net for normal maintenance and 0.28 mill/MJ (1 mill/kW h) net for repair work. The total updated operating cost could thus amount to 0.54 mill/MJ (1.9 mills/kW h) net.

Before concluding the cost analysis, a few remarks should be made on the yearly average production of a geothermal power plant. From the figures available for Italy and partly for other power stations in the world, it can be said that while the load factor (the ratio between the operating hours and total hours of the year) can attain values as high as 98 % (Ricci & Viviani 1970), the plant capacity factor (the ratio between the available power and the installed capacity) generally ranges from 75 to 85 %. The loss of availability is due to accidental failures and, most of all, to lack of steam to operate the plant. Therefore, it seems reasonable to consider the following average values: 8000 h/year and 8500 h/year, respectively, for the condensing and exhausting-to-atmosphere units.

From the preceding analysis of cost components, a conclusion can be drawn regarding the present costs of geothermal energy. They are shown in table 3, where reference is made to the main plants in operation. As table 3 shows, the total cost ranges between 1.4 and 2.5 mills/MJ (5 and 9 U.S. mills/kW h) net. No detailed information is available for the present cost of geothermal energy produced in condensing power plants supplied by steam flashed from mixtures of water and steam, like Wairakei, Otake, etc. Presumably, their present cost should be somewhat higher than the above values.

A comparison can be made between costs of geothermal energy and of that produced by a few conventional plants by reference to the cost figures given by Kaufmann (1970), which can be updated as follows:

power plant	type	capacity MW	plant factor (%)	capital cost in U.S. mills (13.5%) per kW h	per MJ	operating cost in U.S. mills per kW h	per MJ	total cost in U.S. mills per kW h	per MJ
Fall River Mills	hydro	56	60	5.13	1.43	0.63	0.175	5.76	1.6
Humboldt Bay	steam	102	60	4.64	1.29	9.02	2.5	13.66	3.8
Humboldt Bay	nuclear	60	70	8.25	2.29	6.27	1.74	14.52	4.04

Endogenous fluids can be utilized not only for geothermal electric production, but also for other purposes. The exploitation of hot water, which occurs in large parts of all the continents, is also important (Einarsson 1970). Today hot water is principally exploited for domestic heating, particularly in Iceland and Hungary. The Reykjavik Municipal District Heating Service in Iceland now supplies hot water to the houses in Reykjavik at a price of 0.16 U.S. \$/m³. The temperature of the water is about 80 °C and the heat utilized is on the average 176 kJ/kg. This gives an energy price of 0.9 U.S. \$/GJ, which is almost half the price of heat from fuel oil (Palmason & Zöega 1970). The geothermal district heating plant in Sgezed (southern Hungary) is economically a very successful project, actual heating cost; including amortization is

TABLE 3. GENERATING COST OF GEOTHERMOELECTRIC ENERGY

kind of natural steam	superheated steam								wet steam
power plant	with condensing units						with exhausting-to-atmosphere units		
	The Geysers		Larderello		Matsukawa		Larderello		Namafjall
net capacity/MW	2×55	2×28	2×26	1×15†	1×20	2×20	1×15	1×4†	1×3‡
cost of research, drilling and pipelines/10^6 U.S. $	20.53	10.45	7.70	2.22	unknown	unknown	3.78	1.26	0.76
cost of power station/10^6 U.S. $	13.75	7.56	8.84	3.39	7.88§	10.52§	1.42	0.42	0.18
total cost/10^6 U.S. $	34.28	18.01	16.54	5.61	unknown	unknown	5.20	1.68	0.94
total capital cost (13.5%)/10^6 U.S. $	4.63	2.43	2.23	0.76	1.06	1.42	0.70	0.27	0.13
use of plant capacity/h year^{-1}	8000	8000	8000	8000	8000	8000	8500	8500	8500
net annual production — GW h	880	448	416	120	160	320	127.5	34	25.5
net annual production — 10^{15} J	3.17	1.61	1.50	0.43	0.576	1.15	0.46	0.12	0.09
capital cost per MJ (kW h) — fluid/U.S. mills	0.875 (3.15)	0.875 (3.15)	0.69 (2.5)	0.72 (2.6)	unknown	unknown	1.11 (4)	1.39 (5)	1.11 (4)
capital cost per MJ (kW h) — power plant/U.S. mills	0.59 (2.12)	0.63 (2.27)	0.78 (2.8)	1.06 (3.8)	unknown	unknown	0.42 (1.5)	0.47 (1.7)	0.36 (1.3)
capital cost per MJ (kW h) — total/U.S. mills	1.465 (5.27)	1.50 (5.42)	1.47 (5.3)	1.75 (6.3)	1.83 (6.6)	1.22 (4.4)	1.53 (5.5)	1.86 (6.7)	1.47 (5.3)
operating cost per MJ (kW h)/U.S. mills	0.148 (0.53)	0.288 (1.00)	0.55 (2.0)	0.22 (0.8)	0.53 (1.9)	0.288 (1.0)	0.44 (1.6)	0.17 (0.6)	0.14 (0.5)
total cost per MJ (kW h)/U.S. mills	1.61 (5.80)	1.78 (6.42)	2.03 (7.3)	1.97 (7.1)	2.36 (8.5)	1.50 (5.4)	1.97 (7.1)	2.0 (7.2)	1.61 (5.8)

† With remote and radio-control system.   ‡ Pilot station, with a turbine of the simplest possible type.   § Data by Mori 1970 updated.

0.7 U.S. \$/GJ, as compared to the 2.6 U.S. \$/GJ of a coal-fired district heating plant of the same output (Boldizar 1970).

The geothermal plant near the Lake Myvatn (Iceland) for the drying of diatomite has been in use for some years. The exploitation of this product would have involved \$12 per tonne if fuel oil had been used, while the actual cost by using the heat of geothermal fluids is \$2 per tonne.

FUTURE PROSPECTS OF GEOTHERMAL ENERGY

In the countries which at present utilize geothermal energy, research is very active.

In *Italy*, the discovery of steam fields in the Mt Amiata region, in an area where no surface thermal manifestations occur, marks a fundamental turning-point in the research on geothermal fields, and demonstrates that such research in new areas may increasingly be improved by adopting advanced prospecting techniques. The improvements in prospecting techniques makes research even more economical, since the less favourable areas are rapidly located and discarded and costly deep drilling is therefore reduced.

To these aims E.N.E.L. (State Electrical Power Board) and C.N.R. (National Research Council) have launched a joint programme to examine the possibility of increasing the supply of energy from geothermal sources through the discovery of steam fields in new regions (Leardini & Tongiorgi 1968). Already under way are preliminary surveys and prospecting in regions where desk studies had previously indicated the presence of favourable conditions, such as in the areas of Travale-Radicondoli, Monti Volsini, Monti Cimini (see figure 2), etc. At the same time, investigations are carried out in order to improve the basic knowledge of the factors which determine the conditions necessary to the existence of a geothermal field. Based on this policy, careful attention is given in Italy: to reduce the research cost and risk; to extend geothermal investigations to new and wider areas; to obtain higher efficiencies from the fluids exploited in the already known geothermal fields.

A new well with a flow rate of 326 t/h of superheated steam at 184 °C, 1 MPa and 12 % by mass of incondensable gas was put into production, in an area near Travale-Radicondoli (Tuscany) in 1972. A back-pressure unit rated 15 MW has come into operation in this area since July 1973.

A further possibility of increasing geothermal energy production in Italy depends on the replacement of exhausting-to-atmosphere units and the increase of the efficiency of the existing condensing power plants. The new sets installed recently at Larderello (e.g. Ciapica 1970) ensure a fluid consumption of less than 9 kg per net kW h at the station terminals. If an average specific steam consumption for the areas of Larderello and M. Amiata of 10.5 kg of fluid per kW h can be reached, (which seems a reasonable target) a 12 % increase in output could be attained.

In June 1973 the 'Alfina 1' well, drilled in an area near Acquapendente (Monti Volsini), has resulted in a powerful jet of gas and steam. It is planned that deep drilling will proceed in this area in the near future, and a 15 MW unit will be installed with the same characteristics as that recently installed at Travale-Radicondoli.

New sources discovered in the Larderello region are being initially utilized in back-pressure units. For a better utilization of these new sources a 15 MW condensing turbo-generator is being added to the Serrazzano power plant. This unit will come into operation in 1975.

As regards the availability of the primary energy source, it is worth recalling that the total increase of fluid obtained in recent years in the Larderello and M. Amiata regions was achieved only with difficulty, owing to the fact that the limits of production potential have been reached. Therefore, any future and substantial increase in fluid production is closely dependent on the discovery of new geothermal fields in the other Italian regions where research and exploration are in hand.

In the *U.S.A.* research is being carried out in several areas and the investment of considerable capital is foreseen for the future. In the Geysers area, the current schedule calls for startup of units 9 and 10 at the end of this year. The connecting pipelines are being laid and will be ready to deliver steam at the required times; these units will add another 106 MW to the capacity of the existing power plants. Applications for permission to build units 11 and 12 have already been sent to the California Public Utilities Commission. These units are scheduled for construction in 1974 and 1975, respectively. Each one of the latter units will be rated 106 MW net, thus bringing the total installed capacity of the Geysers area up to more than 600 MW. This capacity could produce electric energy sufficient for the needs of a city like San Francisco (Thermal Power 1973).

Geothermal projects have already been started in other areas, like Brady's Hot Springs in Nevada, Klamath Falls in Oregon, Imperial Valley in California, etc.

It was estimated that the Nation's geothermal resources could supply 132 000 MW by 1985 and 395 000 MW by the year 2000 if an intensive programme of research was developed quickly (U.S. Department of Interior 1972).

In *New Zealand* attention is focused on the recently discovered Broadlands field, which has an estimated potential of 120 MW.

In *Mexico* research and drilling on the producing field of Cerro Prieto are in progress. The estimated potential of the field is such as to allow doubling the present 75 MW capacity in the near future. New research programmes are being developed at Pathé, Los Negritos and Ixtlan de los Hervores.

The *Japan* National Natural Resources Committee estimated that it is possible to develop between 30 000 and 50 000 MW of geothermal power in Japan. The Committee reports that the current state of geothermal electric energy development in Japan is as follows: (1) in the Hachimantai area a 10 MW turbine-generator set, which will be in operation by the end of 1973, is now being installed; (2) at Onikobe, in the Tohoku District, an initial 25 MW power plant is being built; (3) at Katsukonda, Japan Metals and Chemical Co. has successfully delineated a new geothermal field and plans to install 200 MW in four sets rated 50 MW each; (4) at Hatchobaru, near Otake, an initial 50 MW geothermal units is scheduled for operation in 1975. A possible increase up to 200 MW is foreseen.

In the *U.S.S.R.* research has been extended to wide regions of the country and, according to Soviet experts, a possibility of step by step increase of approximately 100 MW is foreseen in the near future.

In *El Salvador*, the first research stage at Ahuachapan is practically completed and a first unit rated 30 MW is under construction. It represents part of a power station that is planned to reach 90 MW at least by 1978.

The government of the *Philippines* constituted a panel to negotiate a loan from New Zealand for the development of geothermal energy in Leyte. Preliminary studies showed that the geothermal energy sources in Tongonan, Buravan, and Leyte can generate 100 MW of power sufficient to supply Leyte and Samar in the near future.

In the *South Sulawesi* (*Indonesia*) it is expected that nearly 250 MW can be generated from nine geothermal fields initially. The exploitation of hot waters, which are present in large areas of all the continents, is also important for geothermoelectric production and offers wide prospects for the future. In order to utilize the energy contained in the thermal waters, experiments have been carried out and pilot plants are being designed in many countries of the world.

In Italy an experimental small-capacity ethyl-chloride plant was operated for a short time in 1942 on the Ischia Island using the geothermal fluids there. In these last two years an experimental gravimetric loop has been under construction, working on the principle of the circulation due to the weight difference of two columns of fluids in a closed two-phase (Water-Freon 11) loop. This equipment will use a normal hydraulic turbine rated 1000 kW (Pessina, Rumi, Silvestri & Sotgia 1970).

In the U.S.S.R. experiments with the Freon hot-water plant installed near the Paratunka river are continuing: prospected reserves of thermal waters over the range of temperature from 50 to 200 °C have been tentatively estimated in U.S.S.R. as being over 8×10^6 m³/day (Tikhonov & Dvorov 1970).

In the United States, power plants using pressurized geothermal fluids pumped through a heat exchanger (which utilizes isobutane as a working fluid) are being designed. These plants would harness the hyperthermal waters of the Salton Sea and Brady's Hot Springs geothermal fields, where 50 MW and 10 MW are scheduled respectively for 1978.

In brief, the proposed increases in geothermoelectic capacity in the world in the next years can be summarized as follows:

Italy

Travale-Radicondoli	15 MW		started on 4 July 1973
Serrazzano	15 MW		start in 1975
efficiency improvement of existing units	—	45 MW	in the near future
Alfina	15 MW		start in 1976

U.S.A.

The Geysers units 9 and 10	106 MW		start in 1973
The Geysers units 11 and 12	212 MW		start in 1975
Salton Sea and Brady's Hot Springs	60 MW		start in 1978

New Zealand

Broadlands field	120 MW		start in 1976

Mexico

Cerro Prieto	75 MW		start in 1980

Japan

Hachimantai	10 MW		start in 1974
Onikobe	25 MW		start in 1975
Hatcobaru	50 MW		start in 1975
Hatcobaru	—	150 MW	in the near future

Katsukonda	50 MW		start in 1977
Katsukonda	—	150 MW	in the near future
U.S.S.R.			
Kamchatka	50 MW		start in 1980
Kamchatka	—	50 MW	in the near future
El Salvador			
Ahuachapan	30 MW		start in 1975
Ahuachapan	60 MW		start in 1977
Philippines			
Leyte	30 MW		start in 1978
Leyte	—	70 MW	in the near future
Indonesia			
South Sulawesi	—	250 MW	in the near future
total prospected increase	923 MW	+ 715 MW	

The total geothermal world's capacity of power plants is therefore expected to double and perhaps to triple in the near future, if the scheduled installations will actually be constructed.

Geothermal energy also offers interesting prospects for the so-called 'multiple uses' of natural fluids. The future geothermal development strategy may be directed towards the installation of multi-purpose plants, capable of producing power, drinkable water, valuable minerals, etc.

Finally, it is worth while to recall a recent research programme which is aimed at increasing the supply of geothermal energy. This programme is based on the possibility of extracting thermal energy from the Earth's crust in areas where hot, but essentially dry, rocks can be found at depths less than 6 km. The method of extracting geothermal energy from these areas is really quite simple, since it attempts to reproduce artificially by hydraulic fracturing the heat-transfer mechanism of natural geothermal systems. The water to increase the permeability in this closed-loop system would be maintained in the liquid phase by applying an adequately high pressure (Brown, Smith & Potter 1972).

In conclusion, future geothermal development strategy may be directed first towards the installation of as many as possible electric power stations: since these plants operate on an endogenous energy source, taken from wet and/or dry fields, no foreign exchange costs for importing fuel arise.

In addition to this main use, there are other applications which include the exploitation of terrestrial heat for the heating of houses and other buildings, soil and greenhouses for horticultural purposes, for crop drying, air conditioning and the breeding of animals. Also the installation of multi-purpose plants, capable of simultaneously producing power, drinking water and valuable minerals is being actively considered.

References (Leardini)

Boldizàr, T. 1970 Geothermal energy production from porous sediments in Hungary. *Geothermics* (sp. issue 2), **2**, 99–109.

Brown, D. W., Smith, M. C. & Potter, R. M. 1972 A new method for extracting energy from 'dry' geothermal reservoirs. Los Alamos Scient. Laboratory, 20 Sept. 1972.

Burgassi, P., Ceron, P., Ferrara, C., Sestini, G. & Toro, B. 1970 Geothermal gradient and heat flow in the Radicofani Region (East of Monte Amiata, Italy). *Geothermics* (sp. issue 2), **2**, 443–449.

Cataldi, R., Ceron, P., Di Mario, P. & Leardini, T. 1970 Progress Report on geothermal development in Italy. *Geothermics* (sp. issue 2), **2**, 77–87.

Cataldi, R., Di Mario, P. & Leardini, T. 1973 Application of geothermal energy to the supply of electricity in rural areas. *Geothermics* **2**, 3–16.

Chierici, A. 1961 Planning of a geothermoelectric power plants: technical and economic principles. *U.N. Conf. New Sources of Energy*, Rome 1961. Paper 35/G/62.

Ciapica I. 1970 Present development of turbines for geothermal application. *Geothermics* (sp. issue 2), **2**, 834–838.

Corti, R., Di Mario, P. & Mondolfi, F. 1970 New trends in the planning and design of geothermal power plants. *Geothermics* (sp. issue 2), **2**, 768–779.

Dal Secco, A. 1970 Turbo compressors for geothermal plants. *Geothermics* (sp. issue 2), **2**, 819–833.

Dawson, G. B. & Dickinson, D. J. 1970 Heat flow studies in thermal areas of the North Island of New Zealand. *Geothermics* (sp. issue 2), **2**, 466–473.

Di Mario, P. & Tagliabue, C. 1958 Il telecomando del gruppo trasportabile installato in localita 'S. Andrea' *Rass. 'Larderello'*, n. 3–4 March–April 1958.

Einarsson, S. S. 1970 Utilization of low enthalpy water for space heating industrial agricultural and other uses. *U.N. Symp. Dev. Utiliz. Geothermal Resources*, Pisa, 1970, rapporteurs report, section x.

Facca, G. 1970 General report on the status of world geothermal development. *U.N. Symp. Dev. Utiliz. Geothermal Resources*, Pisa 1970, rapporteur's report, section II.

Facca, G. & Ten Dam, A. 1963 *Geothermal Power Economics*. Florence: Off. grafiche Vallecchi.

Giovannoni, A. 1970 Drilling technology. *U.N. Symp. Dev. Utiliz. Geothermal Resources*, Pisa, 1970. rapporteurs report, section VI.

Guiza, J. 1973 Summary of the U.N. Informal Seminar on development and use of geothermal energy, *Geothermics* **2**, 36.

Hayakawa, M. 1970 Geothermal energy of reservoirs and heat sources in Japan estimated from heat measurements and other. *Geothermics* (sp. issue 2), **2**, 459–465.

J.G.E.A. (Japan Geothermal Energy Association) 1972 Annual information on development and utilization for geothermal energy in Japan, 1971.

Kaufmann A. 1970 The economics of geothermal power in the United States. *Geothermics* (sp. issue 2), **2**, 967–973.

Komagata, S., Iga, H., Nakamura, H. & Minohara, Y. 1970 The status of geothermal utilization in Japan. *Geothermics* (sp. issue 2), **2**, 185–196.

Leardini, T. 1970 Economie de l'energie géothermique. *Geothermics* (sp. issue 2), **2**, 958–966.

Leardini, T. & Tongiorgi, E. 1968 Utilization of geothermal energy in Italy – recent developments in research production and utilization of natural steam resources. *World Power Conf.*, Moscow, August 1968, 30 pp.

Lindal, B. 1970 The use of a natural steam in a diatomite plant. *Geothermics* (sp. issue 2), **2**, 921–924.

Mazzoni, A. 1954 The steam vents of Tuscany and the Larderello plant. *Arti Grafiche Calderini Bologna*, 1954, 170 pp.

McMillan, D. A. Jr. 1970 Economics of the Geysers geothermal field California. *U.N. Symp. Dev. Utiliz. Geothermal Resources*, Pisa, 1970, paper XI/20.

Mori, Y. 1970 Exploitation of Matsukawa Geothermal area. *U.N. Symp. Devel. Utiliz. Geothermal Resources*, Pisa, 1970, paper VII/24.

Moskvicheva, N. B. 1970 Geothermal power plant on the Paratunka River. *U.N. Symp. Dev. Utiliz. Geothermal Resources*, Pisa 1970 paper VIII/13.

National Energy Authority 1973 Recent geothermal development in Iceland. *Geothermics* **2**, 38.

Noguchi, T. 1970 An attempted evaluation of geothermal energy in Japan. *Geothermics* (sp. issue 2), **2**, 474–477.

Palmason, G. & Zoëga, J. 1970 Geothermal energy developments in Iceland, 1960–1969. *Geothemics* **2**, 73–76.

Pessina, S., Rumi, O., Silvestri, M. & Sotgia, G. 1970 Gravimetric loop for the generation of electrical power from low temperature water. *Geothermics* (sp. issue 2), **2**, 901–909.

Ragnars, K., Saemundsson, K. & Benediktsson, S. 1970 Development on the Namafjall northern Iceland. *Geothermics* (sp. issue 2), **2**, 925–935.

Ricci, G. & Viviani, G. 1970 Maintenance operations in geothermal power plants. *Geothermics* (sp. issue 2), **2**, 839–847.

Takung Pao 1972 Geothermal development in China. *Geothermics* **1**, 132.

Ten Dam, A. & Erentöz, C. 1970 Kizildere geothermal field – Western Anatolia. *Geothermics* (sp. issue 2), **2**, 124–129.

Thermal Power Co. 1973 Annual report 1972.

Tikhonov, A. N. & Dvorov, J. M. 1970 Development of research and utilization of geothermal resources in the U.S.S.R. *U.N. Symp. Develop. Utiliz. Geothermal Resources*, Pisa, 1970, paper 1/25.

Tongiorgi, E. 1970 Geothermal systems. *U.N. Symp. Dev. Utiliz. Geothermal Resources*, Pisa 1970, rapporteur's report, section 1.

United Nations 1970 Proceedings of the Symposium on the development and utilization of geothermal resources, Pisa Sept. 1970, v. 1 part 1, v. 2 part 2 (*Geothermics*, sp. issue 2).

U.S. Department of Interior 1972 *Panel of geothermal energy resources assessment of geothermal energy resources*. Washington, D.C.

Zancani, C. 1959 L'utilizzazione del vapore endogeno a forte contenuto di gas. *La Termotecnica*, no. 12, 583–587.

Discussion

Mr N. L. FALCON, F.R.S. (*The Downs, Chiddingfold, Surrey*)

The geological factors necessary for hot water, wet steam or dry steam hydrothermal fields are essentially similar to those required for hydrocarbon fields – reservoir rocks, cap rocks and structural conditions suitable for the migration and entrapment of fluids. The additional requirement of a source of natural heat of great output which is confined to late Tertiary and recent volcanic belts greatly limits the exploration field and also introduces exploration difficulties due to structural and lithological variations. Those who wish for a general introduction to the subject of geothermal energy and its possibilities will find it useful to read no. 12 of the UNESCO Earth Sciences series under that title, Paris 1973. This contains 15 papers written by a team of international experts.

Phil. Trans. R. Soc. Lond. A. **276**, 527–539 (1974)
Printed in Great Britain

ENERGY CONVERSION TECHNOLOGY IN WESTERN EUROPE

Future trends in utilization of coal energy conversion

By L. Grainger

National Coal Board, Hobart House, London, S.W.1

World reserves of fossil fuels are sufficient for many decades of increasing usage. During the next few decades at least, fossil fuels will be much the most important energy source. These fuels should be exploited in a complementary manner.

Coal represents much the largest potential reserve, followed probably by hydrocarbons less easily utilized than those commonly being exploited now. Techniques exist for the conversion of coal into coke and carbons, electricity, gas and substitute oil-feed stock. Improvements in these processes are possible but their large-scale introduction depends on economics.

Where coal burning can meet a requirement (local heat or steam, or electricity generation) fluidized combustion can be the most efficient process; better integration with mining techniques are possible and environmental considerations are favourable.

Fluidized combustion would be a high priority unit in a 'Coalplex' which could have electricity, gas and oil as possible products. The best mix could depend on the value ascribed to the products and this in turn invokes consideration of the overall economics of energy storage, transport and demand flexibility. Looking farther ahead, coal will certainly remain a vital chemical component for various proposed energy systems and will also probably be able to compete as the energy input into conversion schemes.

The technology of coal utilization may also have applications for other fossil fuels.

The world reserves of fossil fuels are sufficient for many decades of increasing usage, but sooner or later exploitable reserves will be insufficient and it will be necessary to adopt conservationist policies. In the case of oil, such policies will begin to be necessary in this decade even taking into account the production of oil from oil shale and tar sands; on the other hand, coal reserves, being perhaps an order of magnitude greater, could support increasing demands at least until the middle of the next century.

When liquid fuels are in short supply, their prices will rise and their use will be confined to those sectors of the market in which their higher prices are justified by their special properties. In practice, this means that liquid fuels become largely confined to transport and the production of chemicals. When the price of liquid fuels is doubled or perhaps trebled relative to coal prices, it will become economic to produce them from coal; this is likely to be the case in the U.S.A. within a few years but will occur later in the U.K. because of the relatively higher costs of coal and the relatively lower costs of oil. A similar relationship will govern the conversion of coal to substitute natural gas.

Until the time when these conversions become economic, the most important outlets for coal in the U.K. will continue to be in the production of iron and steel and in the generation of electricity.

Iron and steel production

Considerable attention is being paid in various parts of the world to the development of processes for the reduction of iron ore and the production of steel using process routes which avoid the manufacture of metallurgical coke and the consequent dependence upon supplies of coking coal. However, it does not appear that developments in the direct route to steel have kept pace with the continued and rapid improvements which have been made over recent years to the blast-furnace route. Considerable advances have been made in the improvement of coke while reducing the dependence on traditional 'prime' coking coals. Much of this has been achieved by close collaboration between the N.C.B. and the B.S.C. At the same time there have been considerable advances in the operation of the blast furnace itself, in the preparation of the ore before charging, in the use of higher blast temperatures and with the injection of fuel directly into the hearth. There is still scope for improvement – for instance, tailoring more closely the properties of the fuel to the requirements of the furnace. This objective would be greatly assisted by the establishment on a more scientific basis of the way in which the properties of the input fuel affect the overall performance of the blast furnace. There is a real need for a concerted effort on a large scale to obtain these data. There is every confidence that as information is acquired about the technical requirements for blast-furnace coke and their economic implications, the coke-making process can be modified suitably within the constraints of available cokes.

In the longer term, direct reduction processes may become established in particular locations where local conditions are suitable. It is worth pointing out, however, that the amount of energy required in direct reduction processes now under development is very little different from the energy requirement of the blast-furnace route. The introduction of direct reduction may therefore change the form in which energy is supplied and possibly the location of steel-making plant but may not affect the total quantity of energy required to produce steel.

Electricity generation

Fossil fuel will continue through the 1980s to play an important role in electricity generation. New processes being developed will improve the overall efficiency of generation, reduce waste and lend themselves to the introduction of improved methods of environmental protection. However, the current status of solid-fuel stations has sometimes been misunderstood in discussions of fuel policy and it is perhaps desirable to restate this from the coal industry's point of view.

The costs of generating electricity depend strongly upon the load factor at which stations are operated. This is particularly important in the case of stations with high capital costs. Figure 1 shows representative annual costs of operating different types of stations at different load factors. This shows that at very low load factors, gas-turbine systems are economic and at very high load factors nuclear power stations can be economic. Over a very wide range of load factors, however, fossil fuel–steam stations can be cheaper to operate than either gas-turbine or nuclear stations. The lower curve of figure 1 shows how the demand for electricity in a typical year would be built up in an ideal situation from the three types of station.

These calculations are based on the assumption that all the capacity would be new, using cost estimates and accountancy procedures similar to those used in published documents. In

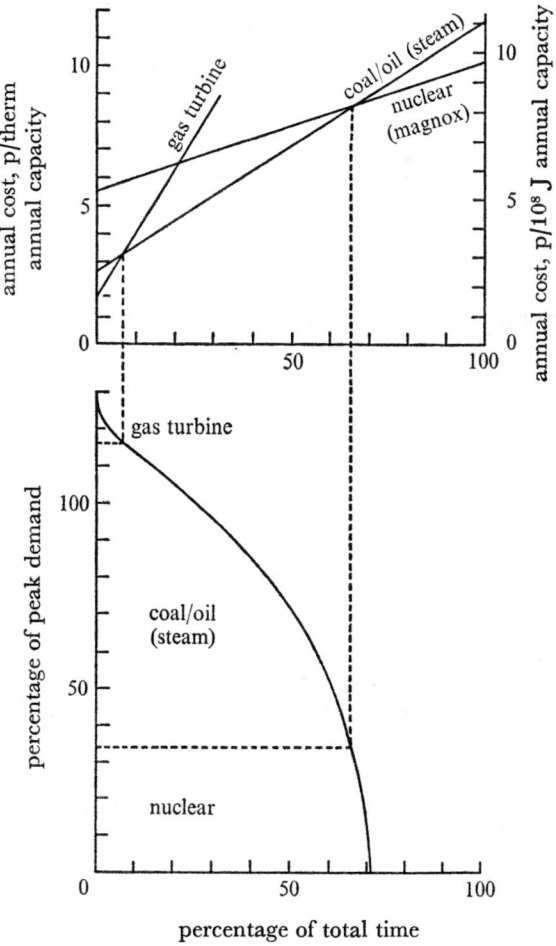

FIGURE 1. Relative future generating costs of electricity.

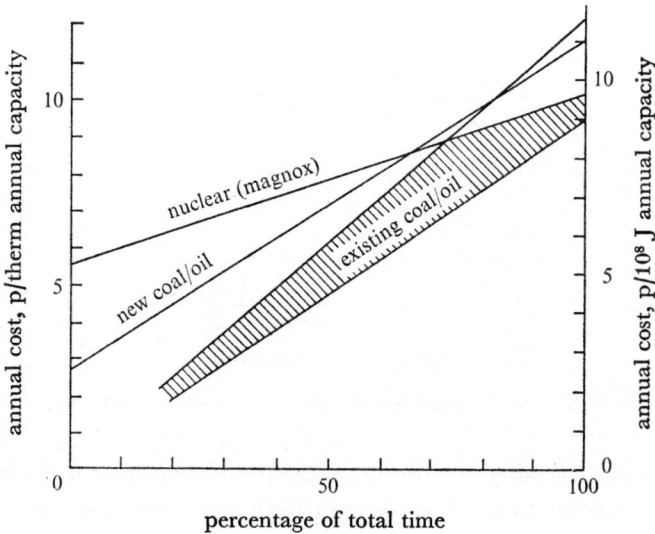

FIGURE 2. Relative future generating costs of electricity.

practice, the system is built up by adding new stations to those already existing. Figure 2 shows, on similar cost bases, that a new nuclear station would not be competitive with existing fossil fuel stations over a very wide range and would become only marginally competitive at load factors approaching the practical limit of availability.

Two new methods of using coal to generate electricity are under active development – fluidized combustion based largely on pioneering work carried out by the N.C.B., and cycles involving gas turbines based on technology for the gasification of coal.

Fluidized combustion has advantages of high heat release rates, high heat transfer rates and thus reduced equipment costs as compared with conventional systems. Because it is smaller and lighter, construction costs would also be reduced. It can be used in pressurised systems, with further advantages. Coals of high and variable ash content can be used and therefore the amount of energy wasted in a conventional system requiring coal preparation can be reduced.

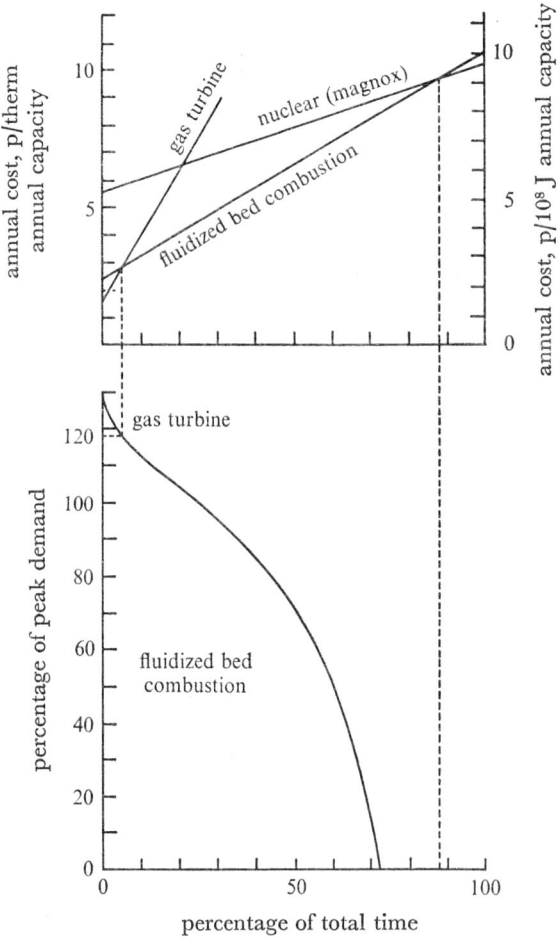

FIGURE 3. Relative future generating costs of electricity.

The system lends itself readily to sulphur absorption techniques. As is shown in figure 3, a fluidized combustion system would be cheaper than published data for nuclear stations at all practical load factors.

In the cycles involving gas turbines, the coal is first reacted with air and steam to produce a low-energy fuel gas from which sulphur can be removed after cooling, by conventional techniques. The clean fuel gas is then burnt in a cycle involving gas turbines and steam turbines. Various arrangements of boilers and turbines are possible, but in each case the objective is to reduce the overall capital cost of the station and to increase the proportion of fuel which is converted to electricity. The extent of these improvements depends on the highest temperatures which industrial gas turbines will accept; at present this is about 850 °C but it is expected to rise progressively following the higher temperatures at which aircraft turbines operate.

The progressive improvement in conversion of coal into electricity is illustrated in figure 4 which shows an increasing proportion of the energy in the coal becoming available to the ultimate consumer.

COAL CONVERSION

The changing energy supply situation has added impetus to the development of new techniques for the conversion of coal to gaseous and liquid products. Work on a modest scale has continued for some years in the U.K. at the Coal Research Establishment. In the U.S.A. following the recent Presidential message on energy, the effort on coal research and development has been stepped up to over $100M a year. Pilot plants are being built to demonstrate the viability of improved technology for coal gasification, to make both low-energy gas and substitute natural gas, for coal liquefaction to make hydrocarbon fuels and chemical precursors, as well as for the development of fluidized combustion. Good links have been established with the American work at government and organizational level and our programmes will take full advantage of American experience.

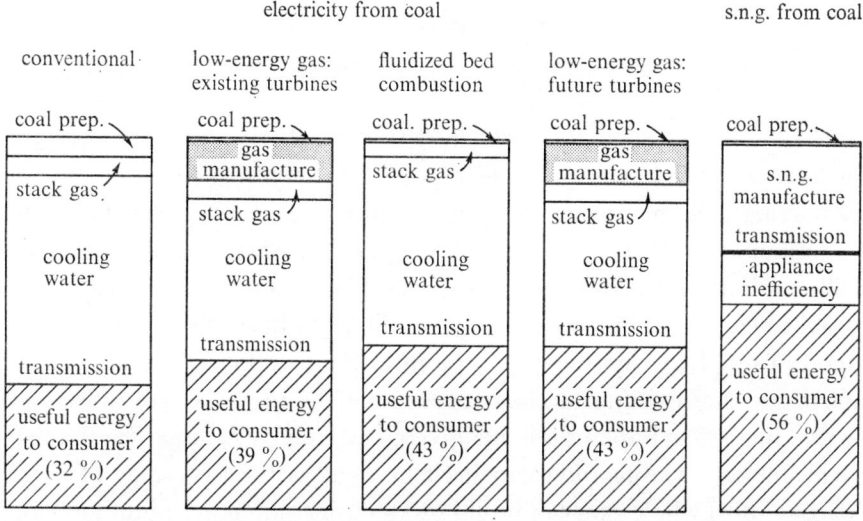

FIGURE 4. Electricity from coal and s.n.g. from coal.

Gasification

Our studies on the economics of the production of substitute natural gas from coal indicate it to be a very much more efficient means of supply of energy from coal than is electricity. The proportion of heat available to the consumer rises to about 56% of the heat originally available in the coal, even after making due allowance for the lower efficiency of the appliance in which it is used. This is illustrated on the right-hand side of figure 4 in comparison with the production of electricity. Figure 5 shows the cost of manufacturing substitute natural gas from coal compared with natural gas from the North Sea and the cost of generating electricity at various load factors. It is apparent that from a national resources point of view the separate development of energy supply systems each involving considerable investment in generating and transmission plant, may not be optimal.

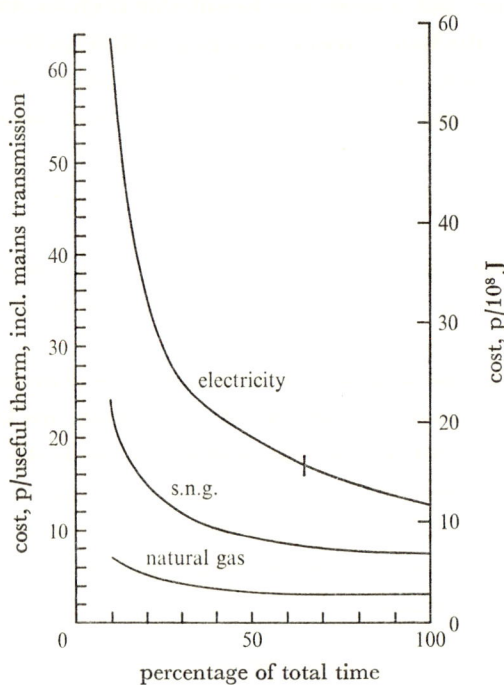

FIGURE 5. Comparative costs of electricity and gas (excluding local distribution).

Coal liquefaction

According to some proponents of nuclear power, the hydrocarbons of the future will be produced by making hydrogen from water by electrolysis and reacting it with carbon dioxide made from limestone. On the contrary, I believe that for a very long time the cheapest source of carbon will continue to be fossil fuel, chiefly coal, and the cheapest way of providing the energy required to convert it into hydrocarbons will be to burn some of it, and the most convenient method of supplying hydrogen will be from the reaction between coal and steam.

This is not to say that existing methods of treating the raw coal are satisfactory; and it is encouraging that increasing impetus is being given, particularly in America, to the development of improved technology. Gasification can itself be used as the first stage of processes for the conversion of coal into a whole range of liquid products including premium fuels and chemicals. The alternative route to liquid products is to attack the coal with solvents. There are basically two variants of the liquefaction process – one uses activated solvents and the other uses hydrogen in the presence of catalysts to break down the coal substance. Instead of using liquid solvents, as in processes being studied in the U.S.A., a new technique is being investigated at the Coal Research Establishment using gases at temperatures and pressures just above their critical point. In this supercritical condition the solvent power is enhanced and solids can be extracted directly into the gas phase and separated from any less-volatile substance which may be present. One advantage of this approach applied to coal is that the problem of separating dissolved coal and mineral matter is considerably eased. It is possible, also, when using supercritical gases, to fractionate the coal extract by reducing the temperature in stages; the first fraction being a low-melting solid and the second a viscous liquid at room temperature.

Pyrolysis

In the process of pyrolysis, coal is subjected to mild thermal treatment in the absence of air or steam. There are three products – heavy oils, gas, and char. By careful control of process conditions, particularly the residence time of coal in the system, it is possible to maximize the liquid production at the expense of gas. The char product may be gasified to synthesis gas or burned to provide process heat. The synthesis gas can be purified and catalytically converted to substitute natural gas.

Pyrolysis thus offers a possibility to produce both oil and gas, with an inherent flexibility which permits variation in the proportions of the products. For example, gas production may be maximized by gasifying the oil, or alternatively the synthesis gas may be converted to liquid fuels.

The flexibility of the pyrolysis approach is such that it is desirable to work in this field in addition to work on direct gasification. The N.C.B. is currently working at Leatherhead under contract to the Cogas Development Corporation on specialized aspects of the Cogas process, which involves pyrolysis.

COALPLEX

I have demonstrated that there are many exciting new methods under development for the conversion of coal into other forms of energy. We have also been studying ways in which these various unit processes can best be combined together to meet the future energy needs of the U.K. Thus the concept of the coalplex has been evolved. The various unit processes are combined together in ways intended to reduce their overall capital costs and to increase the overall

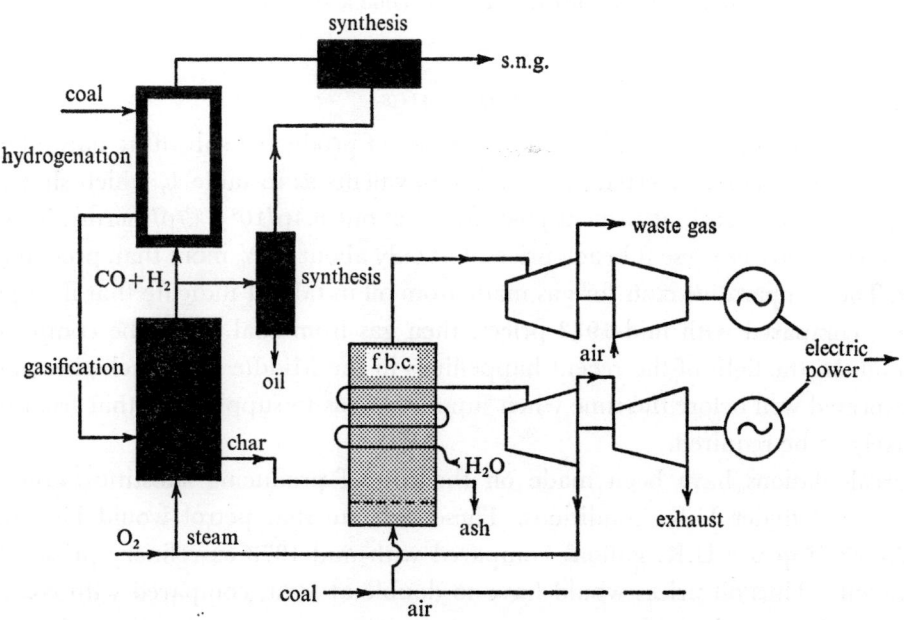

FIGURE 6. Coalplex incorporating gasification and fluidized combustion.

[127]

efficiency while meeting the needs of the energy and raw materials market, taking account of storage and transport costs.

Figure 6 shows a coalplex incorporating gasification and fluidized combustion. The combustion system uses the char produced from the gasification stage.

Figure 7 shows a coalplex based on liquefaction and fluidized combustion. In this case the residue from the main process is first subjected to partial gasification, the remainder being fed to the combustor.

Figure 8 illustrates the use of liquefaction processes as the first stage in the production of specialized carbon products such as fibres and electrode coke.

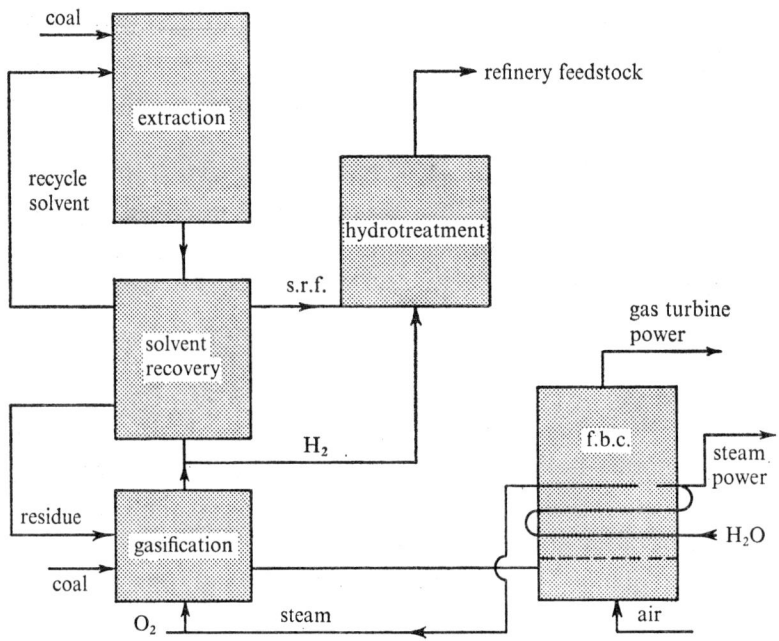

FIGURE 7. Coalplex incorporating liquefaction and fluidized combustion.

ECONOMICS

Various estimates have been published of the cost of producing substitute natural gas in the U.S.A. These have been converted into U.K. equivalents as in table 1, which shows that gas made from coal would cost, at present coal prices, about 6.4p/10^8 J (7p/therm). This is about 3–4 times as expensive as present beach prices, but only about 50 % more than present industrial gas prices. The comparative costs for gas made from oil in table 1 indicate that if oil prices rose by 30–40 % compared with mid-1973 prices, then gas from coal would be competitive with that from oil. In the light of the recent happenings in the Middle East, such rises in oil prices must be expected well before the time when supplies of gas to supplement that from the North Sea are likely to be required.

Similar calculations have been made on the cost of producing substitute crude oil and petrol from coal under U.K. conditions. These indicate that petrol would be produced at about £33/m³ (15p per U.K. gallon) compared with mid-1973 ex-refinery prices of £13/m³ (6p per gallon). Thus oil prices would have to double at least, compared with coal, to make petrol from coal economic.

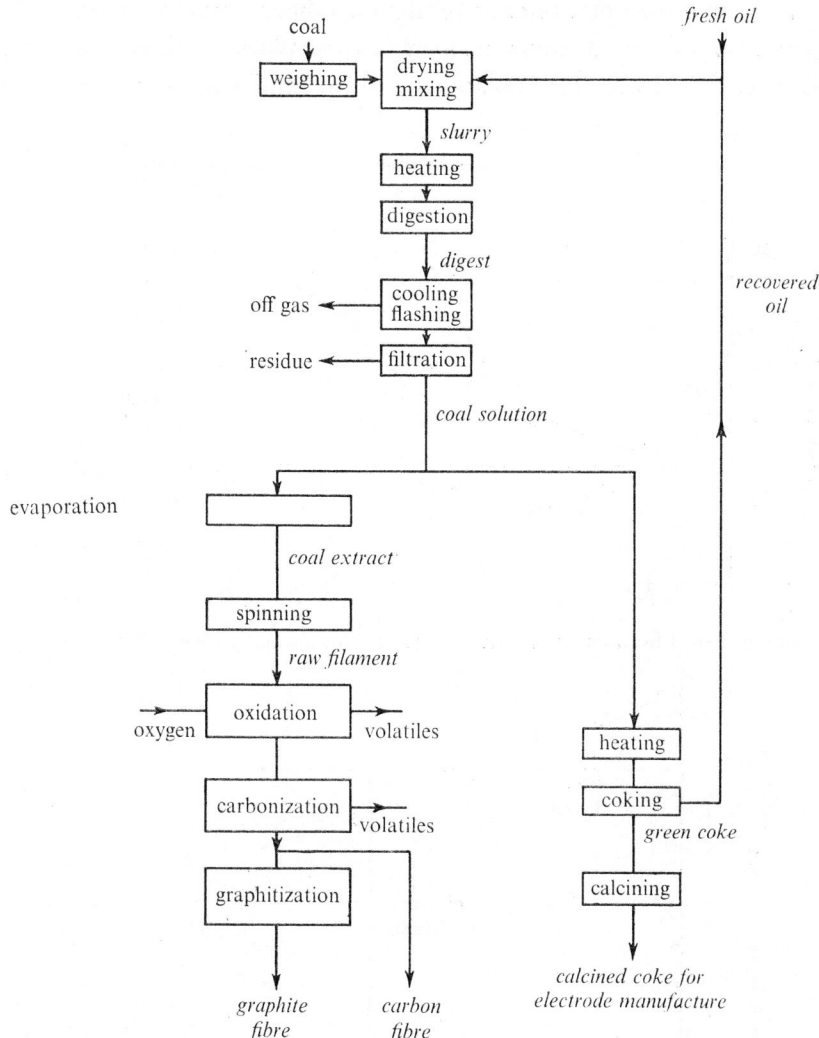

FIGURE 8. Processes for making carbon fibre and calcined coke from coal extract.

TABLE 1. COSTS OF PRODUCING S.N.G. IN THE U.K. – PENCE/10^8 J (PENCE/THERM)

	coal-based system coal at 2.37p/10^8 J (2.5p/therm)	oil-based system	
		oil at 2.37p/10^8 J (2.5p/therm)	oil at 3.22p/10^8 J (3.4p/therm)
capital related costs	1.6 (1.7)	1.33 (1.4)	1.33 (1.4)
processing costs	1.23 (1.3)	0.95 (1.0)	0.95 (1.0)
raw material (oil or coal)	3.7 (3.9)	3.13 (3.3)	3.26 (4.5)
total	6.5 (6.9)	5.4 (5.7)	6.5 (6.9)
by-product credits	0.19 (0.2)	0.19 (0.2)	0.19 (0.2)
gas costs	6.35 (6.7)	5.2 (5.5)	6.35 (6.7)

Figure 9 compares trends in conventional petrol prices with those associated with the productions of petrol from coal.

Economic studies have been made on plants of the Coalplex type and these show the advantages of combining the various unit processes together. Figure 10 is an example of the result of these studies and shows that by combining an oil-producing plant with gas-producing plant, it is possible to show economies. Similar results have been obtained when studying the operation of unit processes together at different load factors. Work continues on investigating various possible ways of improving the advantages to be obtained by combining the unit processes.

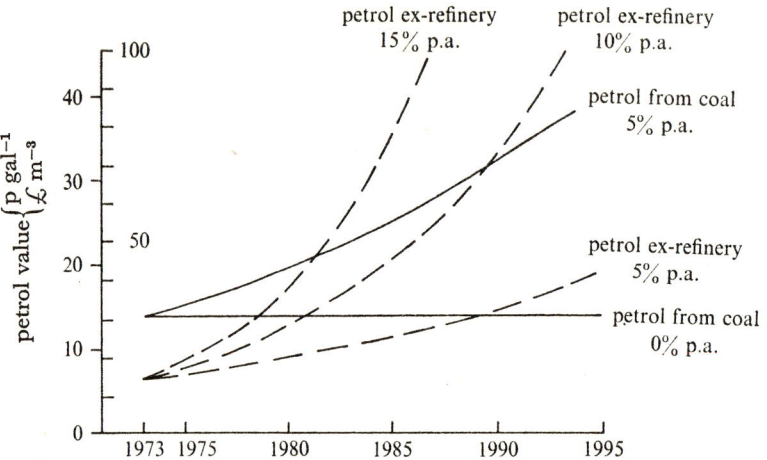

FIGURE 9. Petrol-from-coal processes – effect of cost annual growth rates.

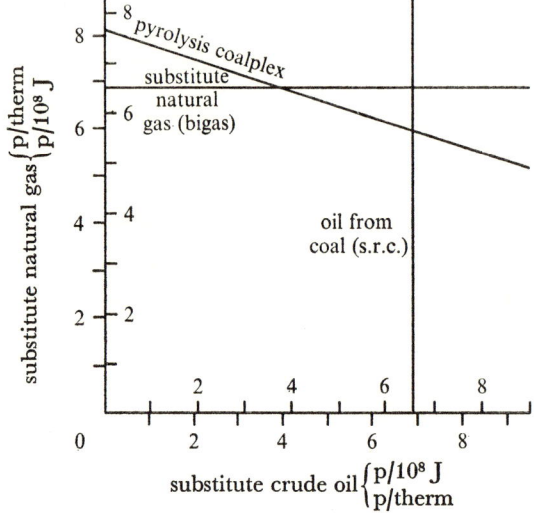

FIGURE 10. Cost advantages of a simple coalplex.

CASE FOR NATIONAL PROGRAMME

The existing research programme at the Coal Research Establishment on the conversion of coal into other forms of energy, while adequate to preserve our familiarity with the technology, needs to be supplemented by operation on pilot plant and on demonstration scales. It is for the latter type of operation that national funds are being sought. An outline of a development programme is shown in table 2. This caters for investment in demonstration and plant-scale operation over a 5-year period. The order in which investment will be made corresponds broadly with that in which the opportunities for coal are likely to arise. Thus the first plant would be concerned with demonstrating fluidized combustion on a 20 MW scale, and would come into operation in 1976.

There would also be a liquefaction unit which would be commissioned early in 1977 to make coal extract freed from mineral matter and suitable for hydrogenation to liquid products. This unit could make the raw material for a special carbons plant.

In 1977 work could also start on the construction of a pilot gasification unit, based on laboratory studies in the preceding three years. The first objective would be to demonstrate an improved process for making low-energy gas. This plant could be integrated with the fluidized combustion unit and would later be used for work on the production of synthesis gas for substitute natural gas and/or hydrogen.

TABLE 2. PROPOSED NATIONAL PROGRAMME OF COAL PROCESSING RESEARCH

£10⁶

	1974/75	1975/76	1976/77	1977/78	1978/79
fluidized combustion unit, 20 MW					
design and specification	0.5	—	—	—	—
construction	0.5	2.0	—	—	—
operation	—	—	0.5	0.5	0.5
demonstration liquefaction unit					
design	0.25	—	—	—	—
construction	—	2.0	1.0	—	—
operation	—	—	0.5	0.5	0.5
gasification unit, s.n.g. low-energy hydrogen					
design	0.2†	0.2†	0.2†	—	—
construction	—	—	—	1.0	1.0
operation	—	—	—	—	0.5
pyrolysis unit					
design ⎱ construction ⎰	0.5	0.5	—	—	—
operation	—	—	0.5	0.5	0.5
analytical studies	0.1	0.1	0.1	0.1	0.1
metallurgical fuels					
special fuel production and testing	0.1	0.2	0.3	0.3	0.3
plant construction	—	—	—	1.0	1.0
total coal processing programme	2.15	5.0	3.1	3.9	4.4

† Programme at C.R.E. supported by E.C.S.C. funds.

The outcome of the present studies of the various pyrolysis systems will be to identify an optimized process which will require demonstration on a reasonable scale. It is proposed that provision should be made for a pilot pyrolysis unit which might be built and operated in 1976/77.

In addition to these plants demonstrating processes for the conversion of coal, it is suggested that it is in the national interest to provide funds for a national programme on the production and testing of special metallurgical fuels, the programme to include large-scale blast furnace trials, and construction of a formed coke demonstration plant.

The overall programme is aimed at the acquisition of know-how in the operation of hardware at the earliest possible time. This will involve a degree of telescoping of the conventional planning of research and development, but the exigence of the situation make it necessary.

The views expressed in this paper are those of the author and not necessarily those of the National Coal Board.

Discussion

Dr I. FELLS (*Department of Chemical Engineering, University of Newcastle upon Tyne*)

Designs for power stations of up to 500 MW based upon fluidized bed combustion of coal were presented some five years ago. It is a pity that at least the 30 MW station proposed then is not under construction now. The lead time for construction of the various plants you have proposed is probably five years at least and it is disappointing that in this country the proposals you have put today still await financial support whilst work is already underway on similar projects in the United States.

The suggestion that we should build demonstration plant is surely not ambitious enough. Problems always arise when scaling up to operational units. We need to build prototype plant for proper economic and engineering evaluation.

SIR KENNETH HUTCHINSON (46–47 *Pall Mall, London S.W.* 1)

A new future has opened up for coal as a raw material for production of refined fuels, and the pioneer work of the late Dr F. J. Dent on coal gasification should now be reconsidered. This work was suspended at a time when the gas industry was forced by market considerations to base its future plans on the manufacture of gas from naphtha, which had become readily available and cheap in the 1960s. The discovery of North Sea gas confirmed this decision. However, we have just seen the United States passing through the era of plenty and heading rapidly for a shortage of oil and gas, not only in this period of political uncertainty but in the long term also.

Dent's most significant discovery was when he showed that methane is synthesized under the conditions of pressure gasification of coal in steam and oxygen, and a hydrogasification stage has been incorporated in the I.G.T. complex in Chicago.

One technological advance in a related field which may prove to be just as significant was the successful operation of a slagging gasifier of the Lurgi type. A pilot unit was brought over from Germany by the Ministry of Power and installed by the Gas Council at their Midland Research Station at Solihull. After complete rebuilding it was shown to be capable of operating without insuperable difficulties at a pressure of 3 MPa, and of outputs of up to four times those of the conventional type. The ash was extracted from the reaction zone in liquid form and quenched in water to give an easily managed frit. In normal Lurgi practice it is essential for the ash to be in a coherent but not in a fused form, and the large volumes of steam which have to be used to keep the ash cool slow up the process, and reduce efficiency.

Meanwhile the Fischer Tropsch process for synthesis of liquid fuel from coal had been studied first at the Fuel Research Station at East Greenwich, and later in an improved form at Warren Spring. Design studies for a coal complex were presented to the Committee on Coal Derivatives (The Wilson Committee) in 1959, but no further development could be recommended at that time, when oil products were so cheap and plentiful. In discussion with Mr Brian Locke the Project Director of N.R.D.C. it seemed to us that a combination of the slagging gasifier as operated at Solihull with the improved Fischer Tropsch synthesis using a slurried catalyst could provide a sound basis for further development. With some further modification it might be possible to produce liquid and gas in about equal proportions, and for both of these there exist almost unlimited markets at high prices. The efficiency of a process designed to produce both liquids and gas could be better than that of processes which produce only one or the other, and there is of course a wide range of options in between.

The pilot plant at Solihull has been demolished and the Lurgi plant at Westfield in Scotland, the last of its kind in Western Europe, seems likely to shut down next year after completing some tests on methane synthesis for American interests. Before a final decision is taken, the possibility that I have outlined might be brought to the attention of those who have already shown interest in the conventional Lurgi process, and have supported further experimental work at Westfield. The slagging gasifier is likely to work best with coals having a low fusion point ash and might be very attractive to certain parts of the United States.

The process outlined here is hardly likely to find application in the United Kingdom in the near future. But if much larger reserves of coal become available than can now be foreseen, or if coal is found at more accessible depths in the North Sea it will be of the greatest value to have experience of a viable process available to us at about the turn of the century. And, in the meantime we might find in this work a valuable outlet for British based technology of a high order.

References (Hutchinson)

Fischer Tropsch *Fuel research 1918–58*. London: H.M.S.O.
Slagging Gasifier 1964 Further experiments with the slagging gasifier. Hebden Lacey and Horsler. *Inst. Gas Eng. Res. Com.* G.S. 112.
Wilson Committee 1960 Report of the Committee on Coal Derivatives. Cmd. 1120.

Phil. Trans. R. Soc. Lond. A. **276**, 541–546 (1974)

Printed in Great Britain

Non-conventional hydrocarbons and future trends in oil utilization in North America and their effect on world supplies

By E. B. Walker, III

Executive Vice President, Gulf Oil Corporation, Pittsburgh, Pennsylvania, U.S.A.

The world's continuously increasing demand for energy and the decreasing supply of conventional hydro-carbons are on a collision course. While there is a range of non-conventional hydrocarbon sources available to draw upon – including oil shale, tar sands, and liquefied and gasified coal – none of these can become an immediate substitute for conventional hydrocarbons in any significant way. Still, these synthetics do have potential between now and 1985 and development must be pursued. The economic reality of synthetic fuels has reached the point where such fuels cannot be considered uncompetitive and unrealistic. The sooner the United States of America can embark on a vigorous development programme for synthetic fuels, the better off will be the nations of Western Europe and other industrial nations dependent on foreign sources of petroleum. Non-conventional hydrocarbons produced in North America probably will not be exported in large quantities but can help to alleviate the pressure for available world petroleum supplies.

Much of the technology for producing synthetic fuels commercially today was developed in Great Britain and Germany several decades ago. With the discovery in recent years of large oil and gas reserves at its doorstep Europe has moved away from man-made forms to the natural or conventional gaseous and liquid hydrocarbons. Across the Atlantic the situation is shifting in the opposite direction, and North America is now placing great emphasis on the extension and improvement of energy-conversion processes.

One fact is becoming abundantly clear to those associated with the energy industry today. The enormous appetite for energy requires the development of the whole spectrum of energy sources as expeditiously as possible. The competition of the future will be *for* fuels, fuels of all kinds, rather than between or among fuels. Energy should be thought of as continuum rather than as a pendulum, swinging from one fuel to another. A shortage in one fuel usually results in increased demand for another.

Petroleum and natural gas satisfied almost all of the growth in post-World War II energy demand, but in the U.S. the oil and gas industry is now at peak capacity. A rapidly rising import requirement in the U.S. will place a strain on the rest of the Western World's petroleum supply and raises some ominous economic, political and national security issues that are important also to western Europe. By mid-1973 there has been the heightening of international political and monetary pressures as a result of petroleum supply. This in turn has heightened an already obvious need to find attractive alternatives. Synthetic fuels, looking much like an ace in the hole, could be one of those alternatives. The price increases emanating from restraints on petroleum production and higher tax revenues make the economics of synthetic fuels more and more attractive. Fortunately the U.S. has ample resources of coal and oil shale, and its neighbour to the north – Canada – has vast tar-sand resources. These provide the base for a massive synthetic hydrocarbon industry in North America.

There is no major synthetic fuels production today, but the potential for growth by 1985 is substantial and development is being pursued. It is, however, unlikely that any North American synthetic fuels will be exported, because domestic requirements already are so great. It should be emphasized, though, that by reducing North American requirements for imports, pressure on conventional world oil supplies will be lessened. In this sense, the development of a synthetic fuels industry in North America is of great importance to the rest of the world. Furthermore, North American self-sufficiency is of vital strategic importance to all the Western World.

With the volatile Middle East atmosphere posing a continuing threat to O.P.E.C. oil supplies, the production of synthetics in both Europe and North America must be viewed quite differently than a few months ago. Because of the dramatic oil price moves in recent weeks set by unilateral action, Europe's own coal and synthetic-fuel industry also may have to be expanded.

Where do things stand now in synthetics in North America?

The following is a brief look at each resource – tar sands, oil shale, and coal – in terms of technology, timetable and economics.

Tar sands are of first importance because there is some commercial production already in Canada and it is the most likely source of substantial supplies of synthetic hydrocarbons in the *near* future.

Tar sands

The largest known tar-sands deposit and the only one of present commercial importance is in the Athabasca region of Alberta, in western Canada. Substantial volumes of tar sands are close enough to the surface to be strip-mined. Even greater volumes, at depths down to 450 m, will require *in situ* recovery techniques not yet perfected. The Alberta Oil and Gas Conservation Board estimates that there are some 47.5×10^9 m³ (300×10^9 barrels) of potential reserves. The Canadian Petroleum Association estimates proved, high-grade, easily recoverable reserves at 10^9 m³ (6.3×10^9 barrels). The technology for tar-sand processing is relatively simple; but there are problems in mining.

There is only one commercial plant in operation – Great Canadian Oil Sands Ltd, started in 1967 – and it is now producing 8000 m³ (50000 barrels)/day by strip-mining the sand and extracting the oil with a hot-water process. This plant is to be expanded.

Syncrude Canada Ltd, a consortium of companies in which Gulf participates, has announced plans for a second plant in this region. It is designed for a capacity of 17000 m³ (108 000 barrels)/day and is estimated to cost more than $900M. A third project has been proposed to the Alberta Energy Board. So after years of a textbook future, tar sands now are moving toward the energy market.

According to the National Petroleum Council's U.S. Energy Outlook Study, which is perhaps the most comprehensive study ever made of U.S. energy, the Athabasca sand deposit is probably the only significant source of commercial production of tar-sand oil for North American markets through 1985. Roughly speaking, it takes about 2 tons of tar sands to produce one barrel of syncrude. Estimates are that the maximum syncrude production by 1985 may be about 0.2×10^6 m³ (1.25×10^6 barrels)/day. A higher level of production may be desirable but the pragmatic problems of design and construction of massive facilities, the availability of capital, and political considerations are indeed limiting.

Oil shale

A second major resource for a North American synthetic fuels industry is the oil shales of the western U.S. Incidentally, oil shale is neither a shale nor does it contain oil; it is a marlstone which contains a bituminous material called kerogen.

The magnitude of this resource is estimated to be on the order of 320×10^9 m³ (2×10^{12} barrels), but only about 8×10^9 m³ (50×10^9 barrels) are both high grade and easily minable. The N.P.C. study identified areas where the deposits are of adequate quality to have potential commercial value today and the N.P.C. estimates that such recoverable reserves, favourably located are about 5.4×10^9 m³ (34×10^9 barrels), based on a yield of 146 l/tonne (35 U.S. gallons per ton) or better.

For a number of years there has been adequate technology for mining and retorting the oil shale but full-scale production will require an enormous materials-handling capability and the resolution of ill-defined but massive environmental problems. Present technology requires handling the oil shale twice, both before and after retorting. Large amounts of water are necessary, and oil shale, depending on the process used, expands quite a bit, so disposal of this waste is a problem. It simply cannot all be put back where it came from. And it takes about 9–12 tonnes of oil shale to make 1 m³ of oil ($1\frac{1}{2}$–2 tons to make 1 barrel).

The oil shale deposits are located in relatively undeveloped and semi-arid areas of the Rocky Mountains. A substantial production capability must be preceded by demonstration plants to determine the best techniques and to minimize environmental problems. The U.S. Department of the Interior is encouraging an industry demonstration programme which should get under way in early 1975, but the requirements for infrastructure, environmental protection, etc., make for very long lead times.

Extensive work on different processes has been done by the U.S. Bureau of Mines and several private companies. The differences are in the design of the retort and the heat transfer mechanism. The Oil Shale Company (Tosco process) has completed operations in a semi-works plant, including several runs of long duration at 900 tonnes a day. In the Tosco process pulverized shale (about 1 cm) is fed continuously into a horizontal kiln. Direct contact with hot ceramic balls heats the shale to retorting conditions. Tosco has worked on the problem of waste disposal and has succeeded in growing grass on this waste.

Gulf is participating with 13 other companies in the Paraho project, which plans to test and demonstrate a vertical-kiln technology that uses more coarsely crushed shale (5–8 cm). The Paraho project, which uses a novel grate that allows continuous operation, has the promise of substantially reducing investment and costs. Both direct and indirect firing methods will be tested. One of the distinct advantages of the Paraho process is that it may reduce the waste problem.

Given the environmental problems of a massive mining operation and the concomitant shale-disposal problem, the development of *in situ* technology is highly desirable for production on a very large scale. A number of companies have been active in *in situ* research, and the U.S. Bureau of Mines is pursuing pilot scale work, but no satisfactory technology is yet indicated.

With the limitation imposed by mining and retorting, the N.P.C. has estimated oil-shale production by 1985 will be in the range of 16 000–120 000 m³ (100 000–750 000 barrels)/day, depending on how great an effort is put into the project. The cost of syncrude at an upgrading plant in Colorado was projected to be from $4 to $7/barrel in terms of 1970 dollars. Until the government defines the lease and environmental requirements, the price can only be approximated.

Coal

By volume of reserves, and as a resource base, coal is the most important potential source of fossil energy in the U.S. today. The N.P.C. has estimated a reserve of 140×10^9 tonnes recoverable in formations of thickness and depth comparable to those being mined currently. The total resource is much greater.

Until recently coal fuelled about half of U.S. electric power generation; it has lost some of its market in the heavily populated U.S. East Coast to low-sulphur oils. Oil is not the long-term solution to electric power generation in the U.S., or to the pollution problems of burning high-sulphur coal. In the absence of technology for stack gas desulphurization and the growing desire for clean fuels in all applications, the vast U.S. coal deposits will have to be converted to some form of non-conventional hydrocarbon.

A joint effort by industry and the Federal Government for research and development in coal conversion is gaining speed. At least two commercial-sized coal-gasification plants have been announced, based on Lurgi technology. These plants are designed to supply pipeline-quality gas, i.e. 37 MJ/m^3 (1000 Btu/ft^3), into the natural-gas distribution system. They are the fore-runner of many more. The low-energy gas from the Lurgi gasifiers will be upgraded by one or several methanation steps under investigation today.

In Westfield, Scotland, Gulf is participating with a number of companies in cooperation with the British Gas Corporation in a methanation demonstration programme that already has achieved some operating success. Because the U.S. has a very large pipeline system and many natural-gas users, this methanation programme is important to the ongoing development work in synthetic fuels.

Nearly all of the technology is in hand, so coal gasification is likely to be the leader in the development of a U.S. synthetics industry. Using the lower heating value sub-bituminous coals of the West, which incidentally are the least expensive to mine, high-energy gas from coal is esti-mated to cost about $1.40/10^9 J ($1.50/10^6 Btu) in the Midwestern or West Coast demand centres.

Several other major demonstration projects are under way; they are construed to be second-generation coal-gasification plants. These are known as the Hygas, Bigas, CO_2 acceptor and Synthane processes, sponsored jointly by the U.S. Government and elements of the energy industry. Some of these processes appear encouraging and attractive but it is doubtful that there will be a full-scale commercial operation much before 1985. The N.P.C. estimates that by 1985 synthetic gas production from coal will range from $14-70 \times 10^9$ m^3 ($500-2500 \times 10^9$ ft^3)/year.

Another coal-conversion project rapidly nearing start-up is the Solvent Refined Coal Pilot Plant at Tacoma, Washington. Early next year Gulf, in cooperation with the Federal Govern-ment's Office of Coal Research, will begin a 45 tonne/day operation at this plant to convert coal of low heating value, high ash, high water content coal to an essentially ash-free, low sulphur, high heating value coal product. It also shows great promise for upgrading the vast bituminous coal deposits east of the Mississippi River; these have high levels of sulphur. Work on a larger-scale demonstration plant can begin perhaps as early as late 1974.

Liquefaction

The technology of *liquefying* coal has trailed that of gasification by many years and only recently has received much attention in the U.S. The F.M.C. Corporation has a 33 tonne/day pilot plant in Princeton, New Jersey, that has produced some oil and a significant proportion of char. Other energy companies are devoting considerable research and development effort to liquefaction.

For many years Gulf has done extensive work on the catalytic hydrodesulphurization of crude and residual oils and presently has in operation a number of licensed units in oil refineries around the world. As an extension of this work, there is some very encouraging bench-scale results for catalytic coal liquefaction (c.c.l.). A feature of the Gulf c.c.l. process is its adaptability to a wide range of coals of varying qualities relative to sulphur and ash content. An expanded and integrated pilot plant is being designed now to provide a basis for a semiworks plant just as soon as practical.

The probable price for coal liquids was estimated by the N.P.C. in 1970 dollars to fall in the range of $38–50/m³ ($6–8/barrel), depending on the cost of feedstocks at the plant. Our best guess is that liquids produced from Western coal have the potential of being very competitive in Midwestern industrial markets. It also appears that coal liquids can compete favourably with synthetic gas from coal, oil from the North American Arctic, and from North American offshore areas not now explored. The attractiveness of coal liquids versus gas is this: a reduced requirement for hydrogen and a lower transportation cost over long distances. Liquids have the added advantage of flexibility in usage and storage.

Given the most favourable circumstances, a large-scale commercial liquefaction plant will not appear on the scene much before the early 1980s. The N.P.C. has estimated a maximum of 108 000 m³ (680 000 barrels)/day by 1985.

Conclusion

By way of summary, it seems clear that from now to 1985 there will be much capital invested in synthetic-fuel development. Although the upper limits of N.P.C. estimates for the synthetic fuels discussed here add up to the equivalent of about 0.6×10^6 m³ (3.75×10^6 barrels)/day in North America by 1985, it is very probable that recent events will accelerate developments. The importance of synthetics is that they will reduce demand on world petroleum supplies, thus aiding Europe and Japan, and will greatly strengthen the Western World strategically.

For this reason, mankind must be awake to new ideas and new possibilities – in even old ideas such as synthetic fuels. The technology and the capability are partly here and partly in the development stage, and the comparative economics are becoming ever more attractive.

And, most importantly, the *time* has come.

Discussion

Mr F. A. SMITH (*Shell Research Limited, Sittingbourne Laboratories, Kent*)

What are the economic objections to gasification of shale?

Mr E. B. WALKER

The principal objection to shale mining is the effect on the environment. At the high gas prices expected in the future, gasification would probably be feasible.

Mr H. B. LOCKE (*National Research Development Corporation, 66–74 Victoria Street, London S.W.1*)

Gasification, and particularly fluidized gasification does indeed seem economically attractive as a means of using the energy in oil shale. Fluidized combustion too, along the lines being developed by Combustion Systems Limited (representing the National Coal Board, British Petroleum Ltd and the National Research Development Corporation), also has some attractions for oil shales and for other low-concentration energy resources as well as fossil fuels as such. The Oil Shale Corporation of Colorado is understood to be considering both approaches. Such technology is also being examined in relation to Athabasca tar sands, high ash graphites in Scandinavia, 70 % ash 'coals' in Czechoslovakia, and other materials elsewhere. In some areas remote from centres of usage it has in the past been assumed that neither gas pipelining nor electric power transmission could be economic: however, modern developments seem likely to enable power and gas transmission to compete in the future with liquid transport in pipes or rail tankers.

The whole family of processes made possible by fluidized combustion, and also the possibilities of oil synthesis already referred to by Sir Kenneth Hutchinson (based upon slagging gasification and slurry synthesis), plus all the various Gas Council developments at present evoking so much interest in the U.S.A. form a considerable body of energy technology. This means that British-based technology should be able to play a most useful part both at home and overseas in energy manipulations in the future. What is now needed is means of connecting the potential users with the development technologists. This is particularly the case with oil synthesis from coal, where plant and staffs still exist at Westfield and elsewhere in the British Gas Corporation, and also at the British Coal Utilisation Research Association at Leatherhead in conjunction with Combustion Systems Limited in London and the National Research Development Corporation.

Phil. Trans. R. Soc. Lond. A. **276**, 547–558 (1974)

Printed in Great Britain

Future trends in gas production and transmission

By D. E. Rooke

British Gas Corporation, London

Increased awareness of environmental factors will heighten the attractiveness of gas as a fuel and the pipeline as a means of delivering energy to the point of use. Demand for gas will be so strong that available supplies will be reserved for premium uses with the result that the design of systems will have to encompass considerable flexibility in load handling.

In addition to imports, new supplies of natural gas will come principally from the deeper waters of the Continental Shelf in the range 200–1000 m and further developments in sub-sea techniques of construction, operation and maintenance will be required. Plants manufacturing natural-gas substitutes from liquid petroleum will be integrated into systems together with large scale storages.

Advances in transmission systems will require better harnessing of known technology rather than a breakthrough into completely new areas of knowledge both in material and equipment. Developments in automatic welding, in-service internal inspection of pipelines and improved data collection and transmission can be expected.

1. Introduction

Within the United Kingdom, the system of gas supply has passed through many evolutionary stages, both technical and organizational, since the first public supply under charter commenced in the early nineteenth century. It has progressed from local distribution on a small scale of unpurified manufactured gas principally for the purposes of illumination to a nationally integrated system of supply under public ownership of a highly refined fuel as a source of heat with virtually no uses as an illuminant. Until a very few years ago, the supply was based upon a hydrogen-rich gas with an energy value of 19.0 MJ/m^3 (500 Btu/ft^3), but with the advent of natural gas from the North Sea the decision was taken to convert the whole country to the direct supply of natural gas at an energy value of a little over 38.0 MJ/m^3 (1000 Btu/ft^3). This process of conversion is now well advanced: by October 1973 over $9\frac{1}{2}$ million premises had been converted, and with the possible exception of one or two isolated pockets where hydrocarbon/air mixtures may still be the basis of supply, all customers will be receiving neat natural gas well before the 1980s.

On the basis of currently contracted reserves of natural gas the U.K. gas industry will be able by the end of this decade to meet 16–17 % of the projected national demand for energy equivalent to approximately 2.1×10^{18} J/annum (5.5×10^9 ft^3/day, 156×10^6 m^3/day). Within this timescale some of the earlier and smaller gasfields to come into production will have started to decline but the introduction of supplies from new fields already under contract will more than make up for the deficiency, resulting in a considerable further growth in the supply system between now and 1980. Depending upon the rate of further discovery and the recovery of associated gas from some offshore oilfields this rate of growth could be exceeded and might well continue beyond the 1980s. There are too many uncertainties to make firm estimates but it is perfectly possible that natural gas will eventually meet up to 20 % of the total energy demand of the country after 1980, say about 2.45×10^{18} J/a. What is certain is that for the whole of the 1980s the characteristics of the supply will be those of natural gas, whether drawn in fact from natural strata or manufactured as a substitute from some carbonaceous raw material.

Since the early 1960s the growth of the gas industry has been rapid. Even during the era of the production of town gas by the steam reforming of naphtha annual increases in load of 10–12 % per annum were recorded regularly; since the introduction of natural gas the rate of annual growth has risen sharply and has touched 30 %. In 1964/65 gas output in Britain was equivalent to 10^9 ft^3/day (28.3×10^6 m^3/day) of natural gas (although the send out was actually of town gas); in the year ended 31 March 1973, it was almost 3×10^9 ft^3/day (84.9×10^6 m^3/day) and the 1967 White Paper target of 4×10^9 ft^3/day (113×10^6 m^3/day) will be met by 1975.

The demand for higher standards of personal comfort in the home, pressure to improve the environment in urban areas under the impetus of the Clean Air Act and a spectacular growth in specialized uses of gas in manufacturing industry have combined to produce this rapid growth. Who can say that these forces are now spent? The 1980s will surely see an acceleration of the social and economic trends of today. Concern for the environment in the widest sense can be expected to grow and legislation against pollution to develop further, automation in industry will increase and man will demand yet higher standards of personal comfort. These trends will combine to maintain a strong demand for natural gas as a convenience fuel, since it is easy to purify, its production leaves no permanent scars on the landscape, it is readily transported in pipelines underground and requires no individual storage at the point of use. It is ash-free, controllable yet flexible in use and the products of combustion, being almost wholly carbon dioxide and water vapour, can be discharged to the atmosphere by a simple flue. The events of recent years and, more particularly, of recent weeks in the Middle East have also served to underline the security value of an indigenous fuel and it seems certain that at competitive prices potential demand could far outstrip supply. In these circumstances available supplies will have to be reserved for the premium uses where resource savings are greatest and this implies that the future supply system has to be designed to accommodate the large seasonal changes in demand which are characteristic of a wholly premium market with its high content of domestic and space-heating load.

A very real problem for the future may be how the selective consumption within industry is to be achieved. Successive governments have seen the need to use the nationalized industries as an economic weapon so that, even in periods of apparently reduced restraint, particular distortions have been imposed upon the pricing structures of some public industries. It would be totally unrealistic to expect that this system is about to change and that the problem of energy allocation is readily soluble by normal market force mechanisms.

2. Sources of gas in the 1980s

Onshore U.K. exploration prospects are not over-encouraging and by the mid-1980s the major prospects on the continental shelf in water depths up to about 150 m should have been explored fully. During the 1980s therefore the major exploration effort will have to be directed to the deeper waters of the continental shelf, say, between 200 and 600 m, or even perhaps as deep as 1000 m. While preliminary exploration of these prospects can be expected to be accomplished by seismic methods closely related to those in use today and, with the addition of dynamic positioning, the larger type of semi-submersible drilling vehicle now coming into use should be capable of accomplishing the drilling, a major change will have to take place in production methods.

(a) Production of natural gas offshore

It can be expected that production wells will be completed on the sea-floor with submarine collection and treatment facilities. Much of the equipment needed, such as well-head cellars, separators and control equipment, is already available in commercial prototype form, although designed primarily for the handling of oil, and there will need to be further development of specialist gas handling equipment, such as dehydration units, heaters, etc. All of the sub-sea units will be equipped with remote control and handling devices for routine operations, including well repairs, but have the capability to accept lock-on diving bells so that technicians can descend to the sea-floor to carry out maintenance and deal with the more awkward operational problems of the oil and gas handling equipment in normal atmospheric conditions. Further engineering development will be necessary before current sub-sea systems and procedures could be considered routine, but there is no reason to suppose that normal commercial pressures will be insufficient to ensure completion of all the detailed development work within the next ten years.

Disposal of the oil produced from these deep waters should not present major new problems since it will be possible to establish oil storage tanks on the sea-bed adjacent to the production facilities and loading into tanker at a floating platform can be visualized. Batch loading of gas, however, although technically possible, cannot readily be economic and submarine pipelines will still be required. A major problem to be solved, therefore, is that of how to lay large-diameter gaslines in such water depths: it seems clear that a straightforward scaling-up of conventional floating barge equipment is unlikely to be successful and the answer probably lies in the development of large semi-submersible lay barges, again equipped with full dynamic positioning. Repair of a deep pipeline in case of damage will also have to be thought out, although the likelihood of damage should be reduced in these deep waters once the pipe has been laid.

In the very nature of things, discoveries in very deep water will be a long way from shore and recompression of the gas *en route* will be necessary. Where possible the most likely scheme would seem to be the collection and movement of the gas by submarine pipeline into an area of water depth of around 200 m, where a platform would be installed to carry the gas compression equipment. Such a platform could be expected to consist of a concrete base section in the form of one or more columns surmounted by a steel superstructure housing the machinery. The superstructure would probably take the form of a dumb barge which could be built, equipped and fully tested in a shipyard before being towed out to location and, after the erection, requiring only final pipe connexions to render the package operational. In this way conventional gas-turbine-driven centrifugal compressors could fulfil the compression duty and operation and maintenance of the essential equipment should not present undue difficulty. It has also been suggested that sea-floor compressors will need to be developed, driven either electrically or hydraulically, the power being generated on the surface and fed down through cables or armoured hoses to the sea-floor. By this means the prime movers could be located on a floating or semi-submersible structure and readily accessible for maintenance with the compressors duplicated so that oceanic vehicles could remove them bodily for servicing ashore. Clearly such a system has a number of added complications, not least that of making and breaking the high-pressure gasline joints and keeping the compressor free of seaborne detritus, but in waters too deep to permit fixed platforms to be built, this solution should be feasible. However, ocean structures

[143]

are developing at such a rate that current depth limitations will not be valid for long and sea-bed compression may well not be developed within the 1980s.

It has further been suggested that gas might be produced far from shore and then either liquefied on the spot and shipped in cryogenic tankers or converted to electricity for trans-mission. The economics of such arrangements would appear to be so unfavourable that they could succeed only where pipeline transportation to an adjacent market is not practicable.

The costs of producing natural gas from deep water will be high but not impossibly so. Neglecting the influence of inflation, which can be assumed to affect all systems equally, it might be hazarded that improved technology will enable the deep gas to be produced at, say, three-quarters the cost of gas developed by platform drilling in 150 m of water.

(b) Load factor correction

Notwithstanding that the discoveries of natural gas over the next few years are sufficient to meet the demand for premium uses, there will be problems in matching conditions of supply to the needs of the customer. As production sources become located in deeper water and in-creasingly far from shore, the penalty for operating on other than base load will become increasingly heavy. On the other hand as the load becomes increasingly of the premium variety the load factor of use will drop. This is a recognizable feature of gas industry planning today, but the problem is expected to grow more acute into the 1980s and some current measures of alleviation may no longer be practicable. Official policy for energy conservation may prevent the deliberate marketing of a proportion of the gas into bulk heat markets on long interruptible type contracts as a method of load factor correction.

(c) Liquefaction for storage

A method of peak shaving which has a particular additional value in improving system security is that of the l.n.g. plant. Natural gas from the pipeline is drawn off at a low rate throughout the year, liquefied and stored at its atmospheric boiling point of $-160\ ^{\circ}\mathrm{C}$ in large heavily insulated tanks, probably of stainless steel, 9 % Ni steel or aluminium, but sometimes in other countries of concrete. At times of peak demand the liquid is regasified, and since a volumetric change of approximately 600 accompanies the change of phase, a moderately sized tank can accommodate a large amount of gas. Thus, a single cylindrical storage tank approximately 42 m diameter and 38 m high would hold 21 000 tonnes of l.n.g., which, on evaporation, would provide $30 \times 10^6\ \mathrm{m}^3$ of natural gas. The whole of this volume can be released in a matter of 4–5 days if required. By siting the installation near the end of a spur main, in addition to providing peak load supplies it can also be used in the event of a transmission system failure to maintain supplies up to the point of failure. The maximum advantage can only be derived from such installations with freedom to site them in close proximity to particular points on the transmission system and if this is frustrated by purely local objections, notwith-standing the agreement of the competent planning authorities, much of the utility of the method will be lost. There is a growing tendency towards objecting to every incursion into the existing environment, however negligible the damage or worthless the *status quo*. If this tendency is not rationalized the 1980s may be a less comfortable era than the promise of the 1970s would indicate.

(d) *Storage of gas under pressure*

The storage of gas under pressure in vessels, whether above or below ground, can be of significance only in a system of local distribution and even 'linepack' – the volume of gas that can be liberated from a high-pressure pipeline when the pressure is dropped – is of little national significance. This is not often appreciated but derives from the fact that, in general, the extra gas is required generally just when the demand is highest and the flow of gas in the transmission line is causing the greatest pressure drop, thus minimizing the further pressure drop available to liberate 'linepack'. It does of course help to cushion the effects of sudden demand changes and may have particular utility in the period immediately after commissioning a new pipeline when the designed load has not fully developed.

In the volumes required to be meaningful to the transmission engineer, the storage of gas under pressure is limited to storage in geological formations, either man-made or naturally occurring. Such storages are likely to play a significant role during the 1980s. One possibility is the storage of gas in caverns leached in deep salt measures, and construction of one such facility by British Gas, near to the coast in the East Riding of Yorkshire, is already under way. In the chosen locality a bed of continuous salt some 150 m thick lies at an average depth of 1800 m, and in this bed storage 'bottles', each of approximately 220 000 m^3 capacity, are to be leached. At this early stage of development it is not possible to define precisely the pressure range over which the storage will operate, that will depend upon the 'creep properties' of the *in situ* salt, but it is expected that about 30×10^6 m^3 of gas will be recoverable over, say, 50 days, at such a pressure that it will pass straight into the adjacent main transmission line. We envisage being able to construct two cavities at one time, taking water from the sea and returning the strong brine; the construction time is necessarily long and a pair of cavities is likely to take about $2\frac{1}{2}$ years to complete. By the early 1980s we would expect to have some 120×10^6 m^3 of seasonal load gas available from this source, and if the development meets with early success other sites will be sought for parallel developments. A small storage with a releasable capacity of only 3×10^5 m^3 and in a shallow salt bed has been operational as a town gas store for more than ten years, demonstrating that the method is sound in principle, but British Gas is now seeking to operate salt cavity storages at greater depth than has been experienced anywhere else in the world and there are some rheological problems.

Investigation has been made of the possibility of storing gas in disused collieries or other mines, but the problems of achieving secure leak-free storage in hard rock which has been subjected to mechanical mining methods are many and this method does not look promising in the U.K. at least. Equally, it has proved difficult to locate aquifers, having all of the required geometric and geologic properties necessary for secure gas storage, of sufficient size to be economically significant and sufficiently remote from potable water sources to be practical. Many potentially interesting prospects in the U.K. have had to be abandoned after intensive study and even the famous Chilcomb Anticline, near Winchester, the cause of so much Parliamentary Debate at one time, would be only marginally attractive in the economic scenario of natural gas. It is more probable that as fields in the southern North Sea begin to deplete, and the costs of recovering natural gas from farther afield and deeper water begin to have a significant impact, that some recharging of depleted offshore fields will be undertaken. Throughout the land mass of Western Europe, however, a number of onshore aquifer storages are in operation already and the number can be expected to grow steadily as suitable sites are discovered.

(e) Manufacture of natural gas substitutes

As a means of providing peak supplies of gas, British Gas will increasingly make use of its range of unique processes for the manufacture of a natural gas substitute. These processes, based upon the pioneering researches of Dr F. J. Dent, F.R.S., in the field of pressure gasification, have already played an important part in the resurgence of the manufactured gas industry in the 1960s, but are likely to reach their full development in a natural gas environment. The novelty of Dent's work lay in his concept of the direct conversion of carbonaceous raw materials to methane, rather than by the classical route of gasification to form synthesis gas followed by methanation, a highly endothermic process followed by a strongly exothermic one, with all the losses inherent in the large-scale heat recovery which becomes inevitable.

The preferred route for s.n.g. production, particularly for peak-load purposes where low load factors of operation and swift changes in load dictate relatively simple plants capable of semi-automatic operation and of low specific capital cost, is by the catalytic rich gas (c.r.g.) process. Lighter petroleum fractions which are capable of being vapourised and purified so as to contain not more than 0.2 parts/10^6 of sulphur can be reacted with steam at about $450\ ^\circ C$ over a highly active catalyst to form a mixture consisting essentially of carbon dioxide and methane and requiring only simple removal of carbon dioxide to provide a satisfactory natural gas substitute. Three former town gas manufacturing plants have been converted to s.n.g. manufacture to provide a full-scale demonstration of different variants of the c.r.g. process which is the basis of the majority of s.n.g. plants now being constructed in the United States in the face of the rapidly developing natural gas shortage. Eighteen streams of plant with a total capacity of a little over $2 \times 10^6\ m^3/h$ have been ordered; of these, two are already in full operation and seven others are nearing completion.

British Gas is now proposing to convert a few more plants to provide additional peak-load substitute natural gas until the new supplies of natural gas from the Frigg and Brent Fields become available. However, in the 1980s it is unlikely that these converted plants will play a major role and plants specially designed and located to meet the specific load conditions at that time would be constructed.

Catalytic gasification cannot be applied directly to the heavier non-vapourizable oils or to crude oil, but in these cases it is possible to hydrogenate the whole feedstock to gaseous hydrocarbons, principally methane but containing some ethane and possibly some higher hydrocarbons as well. The hydrogenation reactions are strongly exothermic and, if secondary cracking is not to occur, careful temperature control is required. With the lighter oils this can be achieved in a simple cylindrical reactor containing a concentric tube into which the main reactants are injected in such a manner as to cause a rapid recycling of product gas. This is the so-called gas recycle hydrogenator (g.r.h.), a brilliant product of Dent's fertile imagination. With heavy and crude oils, although the main mechanism of hydrogenation is similar, the breakdown of the molecules is accompanied by the production of some particulate carbon. Accordingly, the reaction is conducted in a bed of coke particles fluidized with the hydrogenating gas, thus providing nuclei for the agglomeration of the particulate carbon as well as establishing a close degree of temperature control. This fluid bed hydrogenator (f.b.h.) process is currently being operated on the semi-commercial scale in Japan and will most certainly be available for fully commercial use in the 1980s. It is, of course, an inherently more complicated gasification route than c.r.g. Part of the rich gas product from the hydrogenator must be recycled through a steam reformer to provide the hydrogenating gas or alternatively part of the heavy feedstock must itself be gasified at pressure to produce hydrogen, for example, by a partial oxidation process. Not unexpectedly the specific capital cost of the hydrogenator process is higher than that of c.r.g. and the minimum economic scale is also greater, but a wide range of feedstocks right down to at least light crudes can be accepted.

It is impossible at this stage to predict with any accuracy the part which each of these processes will play in the 1980s. If naphtha were to be available at prices reasonably closely related to the cost of crude, then the c.r.g. process alone would be utilized. However, it seems more likely that British Gas might have to contemplate utilizing some of the whole crude that by 1980 will be flowing from oilfields in which it has an equity stake. In these circumstances the choice will lie between using the fluid bed hydrogenator to gasify the whole crude or carrying out a simple distillation to give a fraction suitable for catalytic gasification, a second as feed for a hydrogenator with the residue being used as a source of hydrogen. While the need is principally for peak-load gas it is probable that a solution based upon running a small base-load crude distillation unit and storing the catalytic process feedstock against the time of extreme demand will turn out to be the more viable but all of the detailed technical and economic studies required to elucidate this choice have not yet been conducted.

British Gas also has in principle the capability to make s.n.g. from coal by methanation of carefully purified lean gas derived from, say, a Lurgi gasification plant. Trials of such a process combination are currently being conducted at the Westfield Plant of the Scottish Region of British Gas on behalf of a consortium of U.S. companies led by the Continental Oil Company. While this remains a technical possibility, however, it seems unlikely that a coal based process would be economically viable in U.K. in the 1980s for base-load production of s.n.g., let alone for peak-load purposes. But British Gas intends to keep in close touch with coal gasification technology and hopes to be able to contribute its fund of know-how in this area to cooperative development schemes in exchange for access to the data obtained. By this means a move back into coal gasification could be made at short notice if it became necessary.

3. GAS PIPELINES

(a) Design parameters

In a highly developed countryside like the U.K. the typical pipeline is 750–900 mm in diameter; it operates at a maximum pressure of 6.9 MPa; it is made from a ferritic steel with a very low alloy content, a yield strength of about 460 N/mm^2 (30 tonf/in^2) and a high resistance to both brittle and ductile rip; it is designed to have a hoop stress of not more than 72% of the minimum specified yield, giving a wall thickness in the region 7.5–12.5 mm.

Throughout the 1980s there does not seem much likelihood of a major stretching of these parameters. While there is an economic incentive to go for higher pressures and larger diameters this is offset by worries about security of supply in the event of a failure and resistance from authorities concerned with safety regulations. The use of a higher strength steel is not justified unless it can lead to a pipeline at a lower cost but such estimates as British Gas has made provide, at best, a marginal case. In any case, wall thicknesses ought not to go below about 9 mm in developed countries; this thickness of steel will resist blows from quite a wide variety of digging implements while lesser thicknesses are easily penetrated by quite common tools such as picks.

(It is frequently not appreciated that by far the most common cause of failure on a major pipe-line comes from external interference.)

In remote areas, in stable ground, where there is little chance of pipelines being damaged and where there is a low risk of hazard if there should be a failure, designers will attempt to reap the economic advantages. Pipelines working at pressures up to 10 MPa and even higher in special circumstances will be met. In such remote areas where there is no need for frequent offtakes and interconnexions, the pipeline is subject to only very simple stressing conditions and the design stress will probably be raised to about 80% of the yield.

A diameter of 1200 mm is probably the largest pipe size yet laid in the West, but with the expected growth in gas supply we might expect to see up to about 1500 mm diameter in the 1980s for the transmission of gas in very large bulk from some remote area. Under these conditions there could be an incentive to use higher-strength materials, but this should present no difficulty as weldable steels of much higher strength and toughness than currently in use for line pipe have long been in use in other engineering fields.

The diameter and wall thickness of pipes are somewhat limited by problems of avoiding distortion during handling and by earth loading. The maximum diameter/thickness (D/t) ratio in current use today is about 75 and it has been estimated that a ratio of 120 probably represents an absolute limit even with the development of new machinery – for example, for making field bends. There is already a tendency to buckling in this process and at higher D/t ratios this would become intolerable.

Undersea pipelines are subjected to different conditions and a number of problems are met during laying, particularly when the water depths exceed about 50 m. The pipeline curvature onto the sea-bed has to be controlled in order to prevent a running buckle along the pipe length: this places a major limitation on the D/t ratio. At sea depths of 600–1000 m the water head would collapse most pipes at present in use and for the future even heavier weight pipe, possibly with stiffeners, will be necessary. As wall thickness increases so we can expect pressures to rise to take advantage of the inherent strength of the pipe wall. Pressures up to 20 MPa or more may be used and pipe wall thicknesses will be approaching 40–50 mm but there does not appear to be any requirement for the use of materials of very high strength.

(b) Welding

In all pipelining work increasing use will be made of automatic welding processes. No major problems with weldability of the steel should be encountered, since all of the high-strength, high-toughness steels mentioned have low carbon contents (0.15% maximum but usually below 0.1%), and this is a major determinant of good weldability. One can foresee, therefore, the principal problems being the application to field construction conditions of automatic processes of the metal-arc inert-gas shielded or self-shielding electrode types originally developed for use in other applications. Particularly where heavy wall sections are employed, the emphasis will be on high metal deposition per run to achieve maximum economy. British Gas already has experience of the use of fully automatic welding equipment in the construction of a major pipeline and it can be confirmed that the quality of welding was excellent. A high rate of production was achieved only after a very considerable technical effort in the early stages of the job; there is no intractable scientific problem to be overcome, but further engineering development is necessary.

Particularly for submarine pipelines, use may well be made of the technique of explosives welding which by 1980 should have developed to the stage at which it is only limited by environmental noise considerations. Explosive techniques will employ a sleeve which is welded either to the inside or outside of the carrier pipe. In either case the explosive force will have to be contained by a mandrel which when used on the inside of the pipe could serve as a line-up clamp. A related technique of explosive crimping might also be used under water to quickly shut off leaking oil or gas lines.

(c) Special materials for pipelines

Pipelines have been built in aluminium, stainless steel, reinforced plastics and so on, but all for special purposes, and it does not seem likely that such material will find widespread use for simple gas lines. However, the need for larger-diameter pipes may favour the use of spirally welded pipe, since the ingot size and hence plate width that can be handled in any particular pipe mill places an upper limit on the diameter of longitudinally welded pipe that can be produced. By 1980 systems of producing spiral-weld pipe on site may have been developed to a stage of real practicability and this would then open the way to consideration of the composite spiral pipe. Improved properties could be obtained in large diameter, thin-wall pipes, if they were to be built up from several layers of thin sheet, just as plywood has certain structural benefits over single sheets of wood. Given adequate spiral-pipe technology one could imagine the continuous production of composite pipe by the use of thermosetting resin adhesives between layers of steel strip. Such pipe should be inherently tough and very flexible and, if the process could be adapted to manufacture on a barge offshore, it might be a major factor in overcoming the problems of deep-sea pipelines.

(d) On-line inspection

It is standard practice for British Gas and some of the other leading gas transmission authorities to subject a new high-pressure pipeline to an initial hydraulic test at pressures maintained for 24 h which result in stresses up to and beyond 100% of the minimum specified yield for the pipe in question. Such testing is usually called 'high level' or 'yield' testing. The philosophy is straightforward. All defects grow at a rate dependent, *inter alia*, upon the level of applied stress and the higher the pressure to which the pipe is subjected, the more likely it is that any defects left in the line after construction will grow to significant proportions, open up, and leak even if they do not cause a major failure. Defects left after high-level testing will be very small, often so small as to be insignificant, but certainly giving a good margin for subsequent growth with time under the much lower working stress. It is worth noting that in over 16000 km of pipeline tested in this fashion experience has indicated that 72% of the defects detected were found at stress levels above 80% of the specified minimum yield, with 47% occurring above 100%.

Having gone to so much trouble to ensure the quality of the pipeline as constructed, it would be illogical not to attempt to measure changes in service, and this is now receiving increasing attention. For example, the I.G.E. Code of Practice TD/1 'Steel pipelines for high-pressure gas transmission' dated 1970 seeks to establish requalification of a major high-pressure pipeline every seven years. To do so by repeating a high-level hydraulic test will be time-consuming and expensive. In most circumstances it will be impracticable also since there is rarely sufficient transmission capacity on alternate routes to permit a pipeline to be out of service for several months even at times of low load. There is thus a considerable interest in techniques for inspecting the condition of pipelines remotely and without taking them out of service.

By the 1980s the use of on-line inspection techniques via remote-reading apparatus will have become widely accepted by pipe-line owners. The current simple expedients of checking cathodic protection, line-walking and helicopter surveillance will be backed up by complex internal devices or 'pigs' which will scan the inside of a pipe while the gas is flowing, to locate, interrogate and record defects. The data collected will then be used for quantifying and testing the defects for significance by computer-based machines.

Magnetic devices already exist which have some measure of ability in this direction, but they are in their infancy and require considerable further development. As yet they cannot differentiate between a large defect on the outside surface of the pipeline and a small and insignificant defect on the inside surface, and discrimination is the biggest problem of all. Eddy-current devices are also available and these will be adapted and developed to operate alongside the magnetic devices to ease the problem of discrimination, but more sophisticated techniques are required to inspect the zones around the welds which are more likely to contain important defects than the more easily inspected parts of the pipe wall. Techniques, such as elastic wave propagation (ultrasonics) may be suited to this difficult inspection role and such devices are being actively studied.

On-line inspection has an obvious application in high-density population areas where maximum safety is of paramount importance. Equally, it could be of considerable benefit as a monitor on remote pipelines and undersea lines where repair equipment can often only be deployed at a particular season. Thus, a forewarning of the state of deterioration of a line could enable preventative maintenance to be carried out at the most convenient time.

Much of the basic data about how magnetic fields are changed, and how ultrasonic waves are reflected, by different geometries, does not exist; a programme of physical measurement and developments of theories are needed to support progress. The growing science of fracture mechanics will also provide much data about the significance of the defects once they have been quantified. This is one of the most exciting developments which can be foreseen, but it is also one of the most difficult. However, compared with the task of repairing a pipeline, to eliminate the defective section, the cost of the detection system will be small and there is every economic incentive to progress this matter.

(e) Gas compression

In many existing transmission areas the already large capital investment in gas compression facilities will preclude any major changes in philosophy, and additions to the system are likely to conform largely to present practice. However, the increasing cost and demand for natural gas will put emphasis on the conservation of compressor fuel gas and give incentive to install or modify existing units to give higher overall efficiency.

For ease of maintenance and replacement and to accommodate to rapid load changes and quick starts, British Gas normally uses modified aircraft-type gas-turbine units as prime movers in its gas-compression stations and this trend is likely to continue. Current engines, both the Avon and the Orenda, have overall thermal efficiencies of about 25%. Second generation aircraft turbines such as the Spey and RB 211 have thermal efficiencies of up to 35% and will clearly come into service whenever the larger power of these units can be utilized economically.

A typical single-stage centrifugal compressor has an efficiency of approximately 80% for low compression ratios and this is difficult to improve upon. Axial flow compressors with perhaps variable-geometry blading show promise of efficiencies of the order of 90%. If such compressors can be made robust and reliable, large savings can be made for large volume flows. The problem will again be one of how best to incorporate these large machines into the network without losing flexibility or reducing system security in the event of mechanical breakdown.

The trend to remotely operated unmanned compressor stations will continue and into the 1980s there will be an increasing tendency to make compressor stations fully automatic.

(f) L.n.g. pipelines

The technical feasibility of l.n.g. pipelines has been established but current information suggests that the cost of transmission can only be competitive with a gas line for a transport distance of over 2000 km. The number of routes of this length without a major offtake are not likely to be great and no more than very limited use of this technique can be foreseen for the 1980s. Specific engineering problems remaining to be solved include the design and construction of the insulation casing, dealing with the effects of any flow stoppage, the causes and consequences of line fracture, line vent design, and the problems of filling the line and of starting up. In one case-study it took three years to fill the line initially with liquid; the operational methods put forward in another outline proposal would have resulted in all the liquid evaporating before it reached the end of the pipeline.

4. SOME PRACTICAL PROBLEMS OF TRANSMISSION NETWORKS

(a) Design

Until quite recently the techniques for the design of gas transmission systems have been relatively crude. Only in the mid 1960s did computer techniques for calculating flows in branched networks come into general use and they were limited to non-compressible flows and steady-state conditions and hence were mainly of use for the low-pressure distribution system. At the end of the 1960s computer programs which could calculate transient effects in a network under compressible conditions became available; such calculations now form the basis of operations in the U.K. One of the major problems in design of a complex network becomes the identification of the 'crucial case' among the whole array of possible unsteady-state conditions and further developments in computer technique will be necessary to keep the number of design calculations within reasonable limits as transient analysis becomes the routine design method.

(b) Flow measurement

The basic measurements of gas entering a transmission system are the flow and the energy value of the gas. Flow is measured by a pressure-difference device (usually an orifice) with a number of instruments making subsidiary measurements to permit correction for the effects of pressure, temperature, etc. The accuracy achieved in good installations is about ± 2% on flow and ± 0.5% on pressure, and this level of accuracy is sufficient for day-to-day control purposes. But unless considerable care is taken, accuracies very much worse than this will be met in practice and single effects producing errors of up to 17% have been found in some places.

No national or international standard flowmeter prover exists, nor is one likely to for several years on the scale required by British Gas. The annual turnover of the industry which makes flowmeters is only a small fraction of the cash flow which for British Gas is determined by less than 20 main orifice meters and it seems likely that further advances will depend on the initiative of gas industry itself. The possibility of using techniques such as radioactive tracers, laser/Doppler scanners or critical pressure nozzles as provers is clearly present; at the very least they would ensure consistency but can probably offer a considerable improvement in accuracy as well.

The application of modern design methods to subsidiary instruments such as calorimeters and specific gravity meters will also improve the overall accuracy considerably; these instruments have had no real design attention for several decades, until quite recently.

(c) Flow control

Flow control is increasingly being carried out remotely and small blocks of digital computers can be used for this purpose. For the operation of instrument signalling systems and for the control of smaller items like valves and heaters at offtake points, there has been so far little automation and analogue devices have been used to combine signals from instruments, the final result being entered into the telemetry system via an analogue/digital converter. Increasingly, small digital units will take over these functions and, in addition to arranging telemetry data, they can be expected to carry out simple control functions such as turning heaters on and off, and adjusting valves. Although some manning will still be required for maintenance purposes, the small satellite computer will become an integral part of operations in the 1980s either to give cheaper systems or more often for greater security.

5. POSTSCRIPT

The trends in future gas production and transmission have been examined with specific reference to the United Kingdom, but in a technical sense they are equally valid for the whole of Western Europe. However, the extent to which certain lines will be followed will depend critically upon the pattern of energy development in the country concerned and this will frequently be determined, or at least influenced to a major extent, by political factors. While Britain can look forward to a plentiful supply of natural gas during the 1980s from her own indigenous resources, if these are utilized to meet part of the larger market in Western Europe a decline in availability would soon set in, without in the meantime doing a great deal to ease the problems of the region as a whole. It seems likely therefore that there will not be a common pattern of natural-gas development thoughout Europe, and countries lacking indigenous sources will largely have to depend upon imports from outside Western Europe by gas pipeline or by l.n.g. tanker, supported by manufactured substitutes for peak shaving and to give a measure of security against loss of supply. Such major imports will often pose difficult matters of politics, and in such circumstances a lower rate of growth of gas load should be expected.

Discussion

Mr A. B. Lovins (c/o *Friends of the Earth Ltd, 9 Poland St, London W1V 3DG*)

Do modern gas pipelines have joints and metallurgy appropriate for the transmission of hydrogen instead of methane, or would the design have to be altered to make the pipelines suitable for this purpose?

D. E. Rooke

Yes the materials are satisfactory, but with hydrogen the pipes would have less thermal carrying capacity and more power would be needed to compress the hydrogen.

Phil. Trans. R. Soc. Lond. A. **276**, 559–570 (1974)
Printed in Great Britain

Energy conversion to electricity

By D. Clark
Central Electricity Generating Board, London

[Two plates]

Growing awareness of future problems of supplies of hydrocarbon fuels enhances the importance of nuclear power in meeting continuing growth in the demand for energy, and of electricity as the route for the deployment of nuclear power. Acceleration of the growth of the electricity share of the total energy market and of the substitution of electricity for other fuels will entail the reversal of some of the trends of the past decade in the United Kingdom.

The scope for innovation in the technology of conversion of fossil fuels to electricity will be limited in the United Kingdom by future contraction in investment in fuel-fired generating plant. Uncertainty about primary fuel supplies and prices in the medium term calls for flexibility in fuel use during the transition to a mainly nuclear system. A continuing task is the harmonization of the expansion of electricity production with the preservation of the environment.

The electricity industry differs from the coal, oil and gas industries in being only to a minor extent a producer of energy – from water power. It is essentially a converter of energy. Its product inescapably reflects the costs and reliability of its primary fuel supplies. On an energy content basis electricity is bound to cost more than the primary fuels from which it is generated; it sells on versatility and convenience and efficiency at the point of use. The electricity industry of the U.K. is the biggest buyer of primary fuel in Europe. It has a vested interest in looking critically and impartially at all the primary fuels.

Historically coal was unchallenged in the power stations until the mid-1950s. Then oil began to offer lower costs and more reliable supplies in the immediate future; and nuclear power for the longer term. Diversification began to take effect in the late 1950s; stimulated by government in the case of nuclear power and alternately stimulated and impeded in the case of oil.

Figure 1 shows, on the left, the C.E.G.B. generating-plant mix early in the present decade. The system can produce electricity from four primary sources. But it is still predominantly a coal-fired system. European electricity industries collectively have a rather smaller proportion of nuclear plant and a considerably bigger proportion of oil-fired plant.

However, in the worst period of electricity rationing during the coal strike of 1972, when C.E.G.B. electricity production was cut by a third, just about half of the reduced output came from nuclear, oil and gas-fired generation.

Because of long manufacturing times much of the plant for installation in the present decade is already committed. By the early 1980s the C.E.G.B. plant mix will look something like the right-hand column. Users of electricity will be better cushioned than at any previous time against interruptions in supplies of any one of the primary fuels.

We naturally tend to think of indigenous coal as being more secure than imported oil. But continuity of supply does not depend just on which country controls the reserves. It depends on people; on the will of producers and workers to maintain a reliable flow of fuel. Over the past 15 years electricity production has been hit harder by coal stoppages than oil stoppages. Short-term interruptions to either must be expected.

Adjusting the mix of plant to match a changing fuel situation is a slow process, because it takes a long time to get new plant manufactured or existing plant converted. What is even more valuable than diversity of plant by itself is to have flexibility to change the fuel mix at short notice. The British electricity system now has a growing capability to do this. It derives from, three factors: the diversity of generating plant already described, the strength of the supergrid, and the unified control of plant loading.

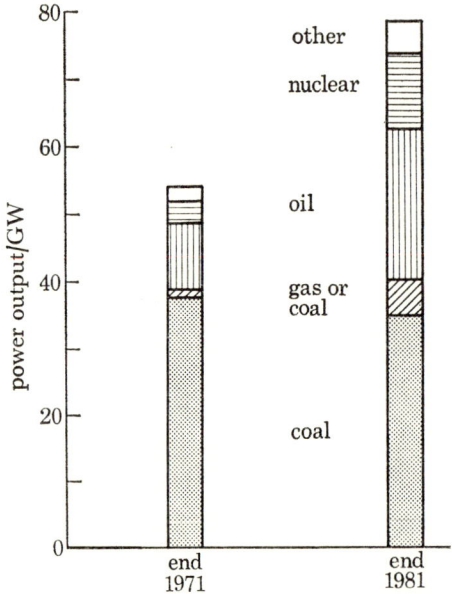

FIGURE 1. C.E.G.B. plant 1971 and 1981.

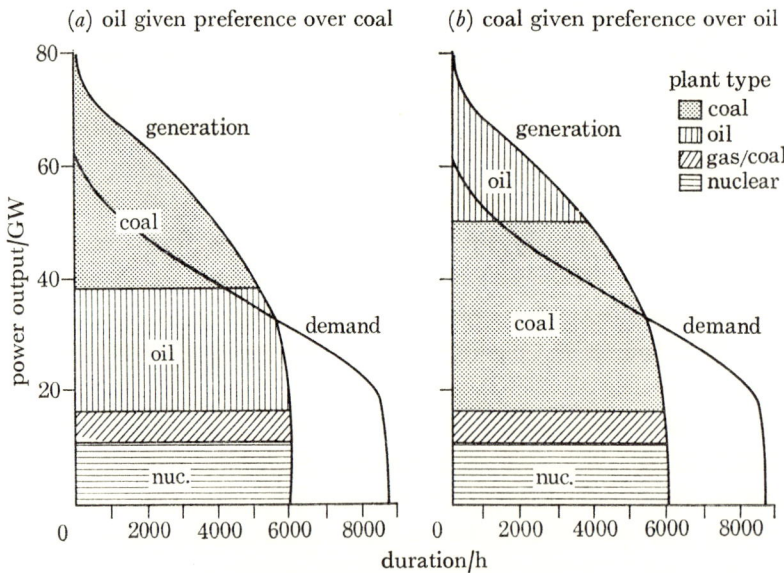

FIGURE 2. Simplified diagram of C.E.G.B. plant loading in 1981.

Electricity cannot be stored so generation must take place round the clock; and there must be enough generating plant to meet the peak demand. For most of the time electricity consumers take less than their peak demand, and when they do a choice can be made as to which plant to run. Plant is normally put on and off load in order of merit in terms of operating costs.

[154]

Figure 2 shows in simplified form the equivalent hours of full load operation of plant, as predicted for 1981/2. The plant with lowest estimated operating cost is stacked at the bottom and runs for the longest periods at full load. In 1981, nuclear plant will still be well within the high load factor area. The left diagram is on the basis of oil being given preference, either on account of price or availability; and the coal-fired plant running for shorter hours. The right-hand diagram shows coal given preference; and the oil plant running for shorter hours. The shaded areas show broadly the potential for producing electricity from each fuel.

Geographically the big oil-fired power stations are on the coast, near refineries. The big coal-fired stations are mostly inland, in the East Midlands and Yorkshire. A switch of preference means big changes in the power flows over the supergrid, but it has been designed with the capacity to handle them. It is this capacity that makes it possible to get substantial flexibility of fuel use from a system made up mainly of single-fuel power stations; as well as securing the classic economies of an electricity grid system.

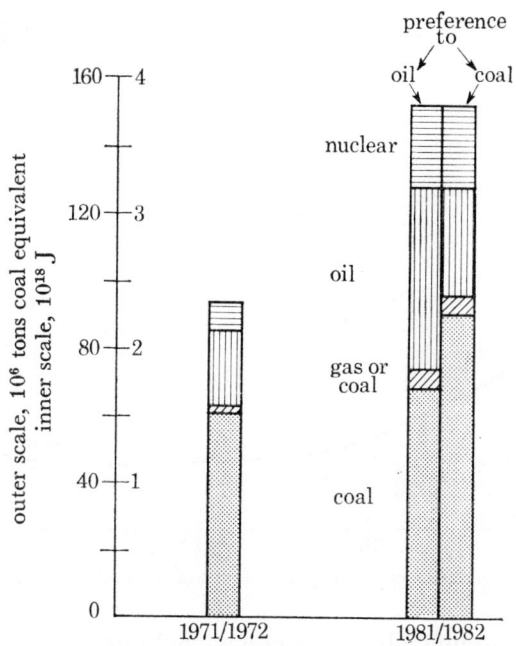

FIGURE 3. C.E.G.B. fuel mix 1971 and 1981.

Figure 3 shows the mix of fuel consumption that results from the plant shown on figure 1. On the left, the actual mix early in the present decade is shown; on the right, the estimated range of possibilities for the beginning of the 1980s; left side, oil given preference; right side, coal given preference. At the present time the range of flexibility as between coal and oil is 2.5–4.0×10^{17} J. By 1981/2 it will be at least 5.0×10^{17} J. Using existing and committed coal-fired plant the C.E.G.B. stations will be able to burn over 90×10^6 tonnes of coal a year if availability and price are right. Scottish stations would add roughly 10% to that capability.

This means that as well as having the largest coal industry in Europe, Britain has an electricity system more than capable of absorbing all the indigenous coal that is likely to be available, at least into the early 1980s. Note also that it is an electricity system still dependent on coal for, as a minimum, half of its fossil fuel requirements.

The foregoing describes what the electricity industry is doing to diversify its fuel supplies;

and, as a major fuel user, to create flexibility to respond to unforeseen events. This is the setting in which energy conversion to electricity in the 1980s has to be seen.

In the years between now and 1980 world fossil fuel supplies may well be difficult – at least intermittently. Conceivably, over-reaction could cause a swing back to surplus; but there is wide agreement that other energy supplies need to be expanded in the 1980s. Tidal power and geothermal power will be limited and local. Solar power cannot be counted on for much in the 1980s. Nuclear power, fission power, is the main hope. Figure 2 shows how much scope there is for additional nuclear plant to operate at high load factors in the U.K. In its early days nuclear power had to face falling real costs of fossil fuels, but for the future the position is likely to be reversed. It is quite possible that nuclear power will become economic and practicable for lower load factors by the time these are imposed by the system.

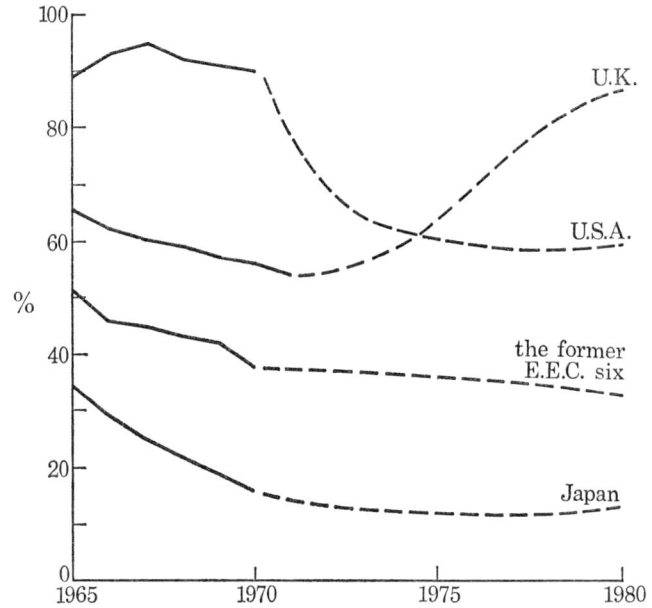

FIGURE 4. Indigenous energy as percentages of primary energy demand.

Figure 4 illustrates the point made by the Secretary of State, namely that in the U.K. North Sea oil will lead to a rising trend towards oil self-sufficiency and towards overall energy self-sufficiency. The build-up portrayed for the U.K. may prove to be a little optimisitc, but even allowing for that, it is in sharp contrast to the prospects for other major industrial countries. The electricity industry looks on this prospect as supporting the continued construction of a limited number of oil-fired power stations in the U.K., but only to bridge the capacity gap before a renewed programme of nuclear construction can take effect. By the middle 1980s nuclear power ought to be providing the whole of the increase in energy for electricity production; and be helping to sustain a high level of energy self-sufficiency in the U.K.

Figure 5 shows the cumulative ordering of nuclear plant in recent years. There are two main reasons for the hiatus in the U.K. since 1969. First, the low rate of electricity growth which meant that very little new plant of any kind was wanted. The second factor was the discouraging experience with the contracts for the a.g.r. stations that were ordered between 1965 and 1969. With rather better electricity growth the slack in plant on order is largely taken up and there is room for substantial nuclear ordering for the 1980s.

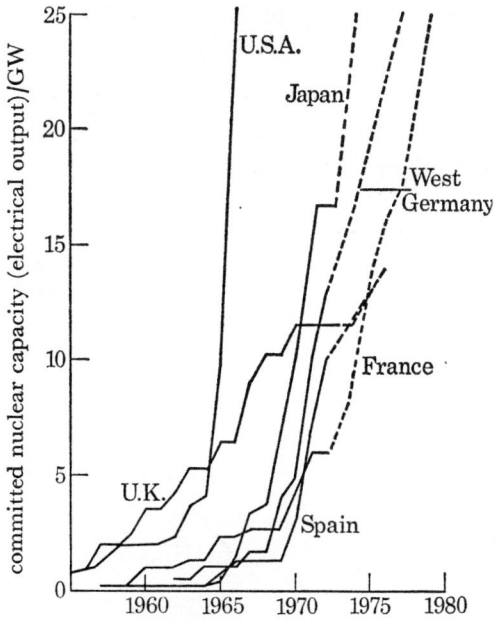

FIGURE 5. Cumulative orders for nuclear plant.

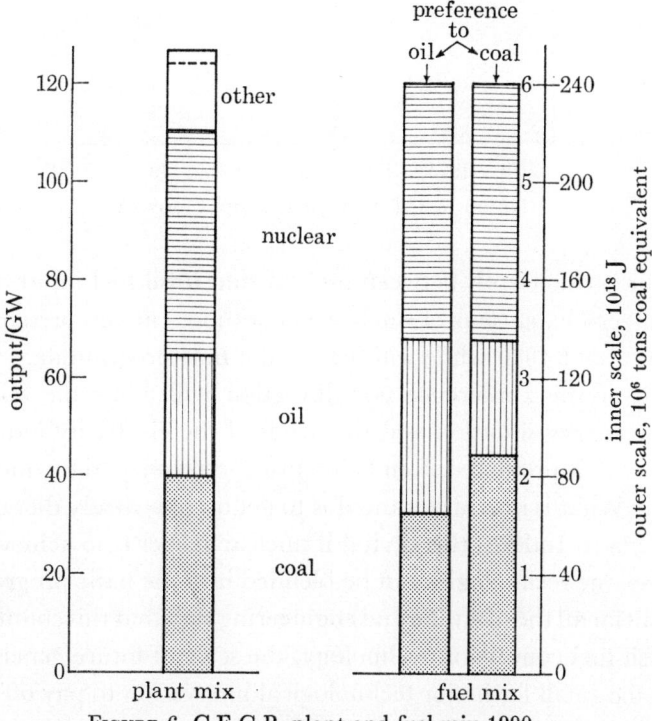

FIGURE 6. C.E.G.B. plant and fuel mix 1990.

If we assume that in the 1980s electricity growth will be at a rate of 5% per annum in maximum demand, then the capacity of generating plant to be added to the C.E.G.B. system in that decade would be about 55 GW. Given an early choice of reactor system some three-quarters of that might be nuclear plant, coming into service at a rate of about 2 GW per annum in the early 1980s and building up to between 5 and 6 GW per annum in the second half of the decade. The remaining one-quarter would include both peak load plant and fossil fuel steam-plant, the choice between coal and oil depending on the trend of supplies.

Figure 6 shows on the left the resultant mix of C.E.G.B. generating plant at the end of the 1980s and on the right the resultant mix of fuels. The assumed split between coal and oil is very tentative and again there is considerable flexibility at the interface.

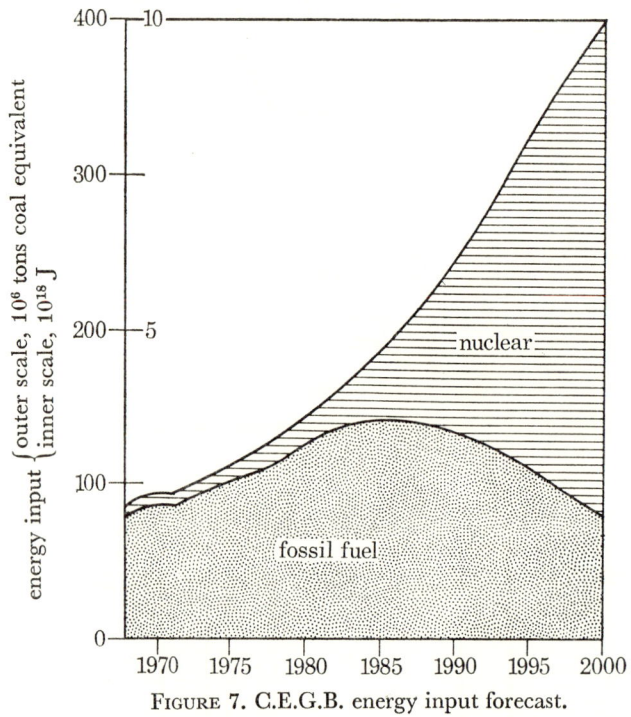

FIGURE 7. C.E.G.B. energy input forecast.

Figure 7 shows the possible split between nuclear and fossil fuel input over a longer period. Note the periods of zero expansion of nuclear generation – at the present time while waiting for the a.g.rs and at about 1980 while waiting for the new programme.

Neither my remit at this conference nor discretion will allow me to digress into reactor choice. But both the energy situation and the needs of the electricity industry make the same demand. Nuclear power is promoted from being just the savoury to the meal to being the main course for the future. What is urgently wanted is to get on to a steady diet of good, well-proven, 'bread and butter' plant. Indeed, this is vital if nuclear power is to achieve its potential in the 1980s. More venturesome technologies can be blended into the basic programme later on. Such a programme will call for all the scientific and engineering skill that this country can bring to bear.

In the case of fossil-fuel conversion technology, the scale of future generating plant ordering in the U.K. will be too small for major technological innovation to pay off – at least in the case of steam plant. Steam turbines suited to water reactors would be essentially adaptations of designs already existing or under development. For other reactors with higher steam conditions some modifications to past practice are probable, such as the use of steam-to-steam reheat in order to simplify the nuclear steam supply system. Unit sizes roughly double the present 660 MW are also probable in the 1980s, though if 1300 MW reactors are built at an early stage they will probably each have two 660 MW turbo-generator sets.

Peak load plant, though a small proportion of the total, is likely to be installed on an expanding

FIGURE 8. Ffestiniog pumped storage station.

FIGURE 9. Trawsfynydd nuclear power station.

FIGURE 10. Sizewell nuclear power station.
FIGURE 11. West Burton power station.

scale. Gas-turbine developments will probably continue to lean heavily on the much larger investment for aircraft propulsion.

The pumped-storage version of energy storage is particularly valuable because hydraulic machines give a faster and more assured response to instant demands for more power generation. Other forms of energy storage will come in for increasing attention because, towards the end of the 1980s, low-cost surplus nuclear generation will become available on a sufficient number of nights of the year to improve storage economics.

The environmental implications of energy growth are as important as the technology. The electricity industry of the U.K. has done more to harmonize the expansion of electricity production with protection of the enrivonment than any other electricity industry that I know. Great care is given to the planning, siting and design of every new power station. Decent appearance of a plant in its setting is a prime requirement. An ugly or untidy plant creates hostility, and the suspicion that it pollutes the air and water as well as the scene.

Figures 8–11 show examples of British power stations. Figure 11 illustrates a problem at inland sites, namely that the cooling towers bulk larger in the landscape than the power stations themselves. And inland sites are unavoidable because many stretches of the remaining undeveloped coastline are among the areas most worthy of preservation.

To ameliorate this the C.E.G.B.'s research and engineering departments have developed a new fan-assisted design of cooling tower. The first will be used at the Ince 'B' oil-fired power station now under construction. One tower will cool 1000 MW of plant, which previously required three or four natural-draught towers having the same chimney dimensions. Two towers will suffice for a station the size of West Burton. Later it is hoped to develop the new type of tower to 2000 MW capacity. A more intractable problem with cooling towers is the appearance and persistence of the vapour plumes in humid weather. A so-called 'dry' cooling tower has been in use at Rugeley Power Station for many years. There is no vapour plume in that case, but the tower is much bigger and more expensive than a wet tower. A study is being made of the possibility of a hybrid tower which might reduce the vapour plume without unduly inflating bulk and cost.

In the case of coastal and estuarine locations, experience and research has shown that with careful design very large quantities of warm water can be discharged without harm to the ecology.

Figure 12 illustrates a topical example. Recently a newspaper magazine published an account of 'a huge settlement of ten million oysters found in the Solent'. This figure shows a contour map of the density of the oysters on the seabed. Several factors are advanced by the biologists to account for this sudden occurrence of oysters in Stanswood Bay. In addition there is a fascinating coincidence. Fawley power station is one of the biggest in the country and the tunnels discharging the warm water terminate right where the oyster bed has appeared. The big spatfall that started the new bed is believed to have taken place in the summer of 1970. Fawley Power Station began effective operation at the end of 1969. Evidently not all change is for the worse.

A continuing environmental objective is to limit the proliferation of power stations by putting as much capacity on each site as the technology of the day permits. In 1960 the biggest output from a single site in the U.K. was 1000 MW. Today it is 2500 MW. Sites suitable for over 4000 MW have been identified. A study is being made of the problems of 10 000 MW groupings, and how to solve them. At the other end of the size spectrum, enough is known about the environmental aspects of gas-turbine stations to allow them to be located close to load centres.

[159]

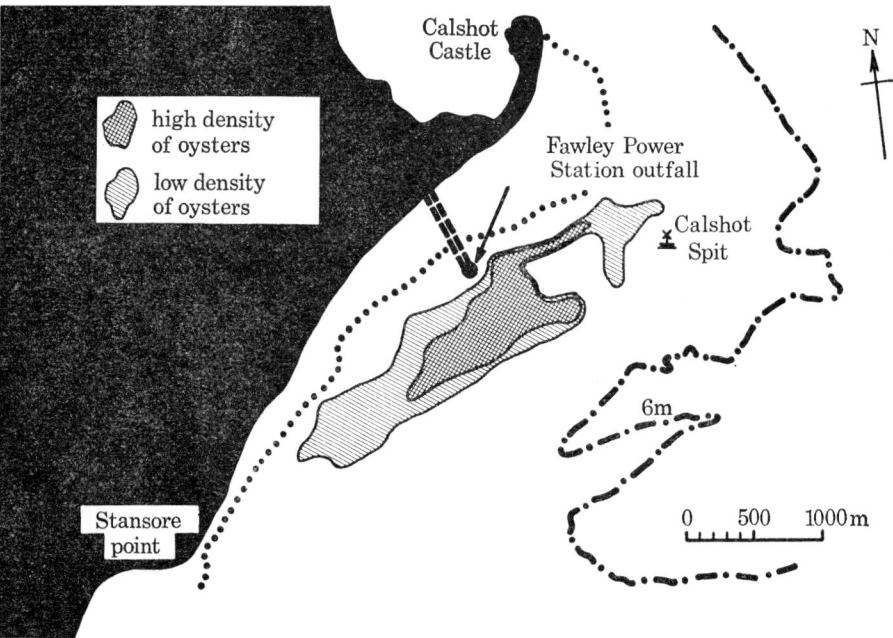

FIGURE 12. Stanswood Bay oyster bed.

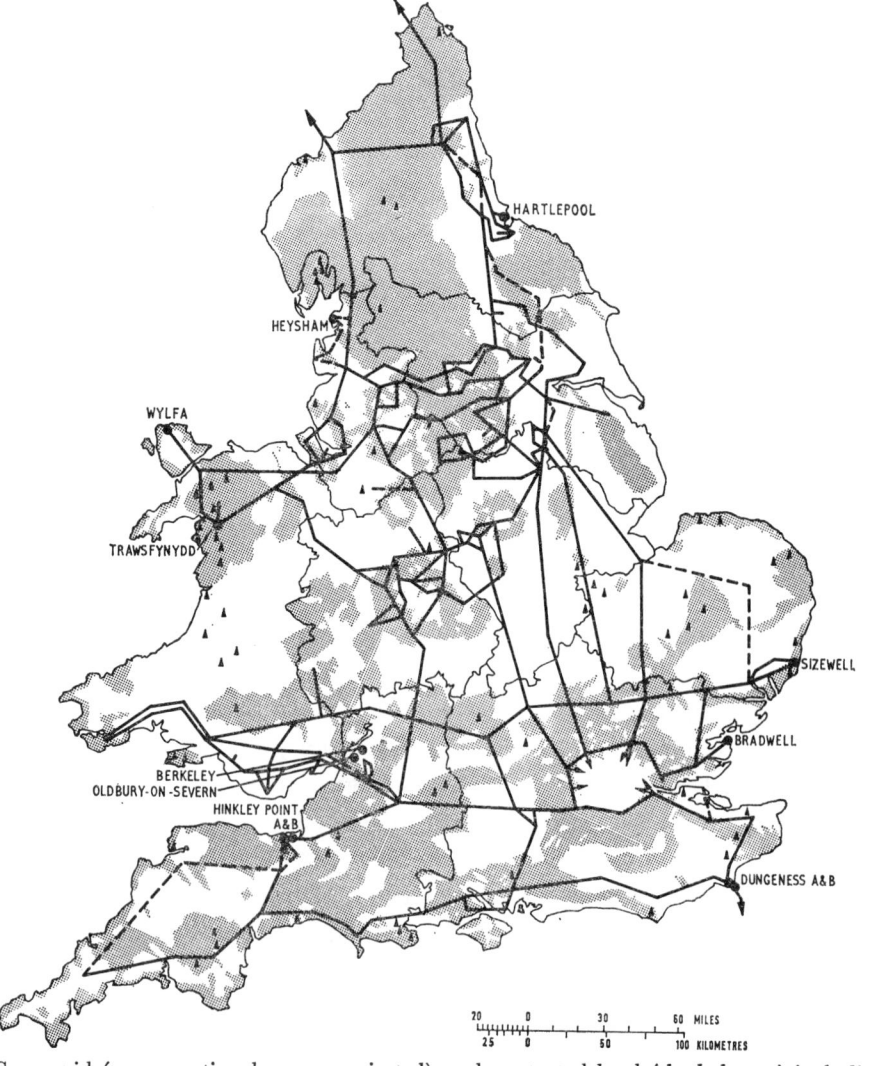

FIGURE 13. Supergrid (——, operational; – – –, projected) and protected land (shaded area) including National
Parks, Areas of Outstanding Natural Beauty, Areas of High Landscape Value, Forest Parks, and Green
Belts (approved and under consideration). ▲, National Nature Reserve; ●, nuclear power station sites.
Standard region boundaries are also shown.

Nuclear generation eases one problem because the transport of fresh fuel to the stations and of spent fuel to the reprocessing plants is on a small scale compared with fossil fuels. Remote siting for safety reasons is apt to be objectionable environmentally, and it is now largely a psychological point. The real safeguard for any nation needing nuclear power is a rigorous standard of integrity in the design, manufacture, operation and maintenance of its nuclear installations.

Figure 13 shows the supergrid system in England and Wales superimposed on a map of the areas which are specially valued environmentally. It is a multi-purpose system; it interconnects all the major power stations and gives bulk supplies to the Area Electricity Boards for distribution. At 1960 the capacity of the biggest line was 1000 MW at 275 kV. Today it is 3500 MW at 400 kV. Means are being explored of getting 5000 MW on one 400 kV line in the future. The main framework of the supergrid is now expected to be adequate through the 1980s. Local additions will be required, mainly to connect new power stations. Research into higher voltages is well in hand in case of need. A direct current cable link is being brought into service within the a.c. network to gain experience of costs, problems and performance.

All in all, there are good grounds for believing that the electricity expansion of the 1980s can be harmonized with effective care for the environment.

I would like to finish with a speculation about energy use to the end of the century, and a possible question for discussion.

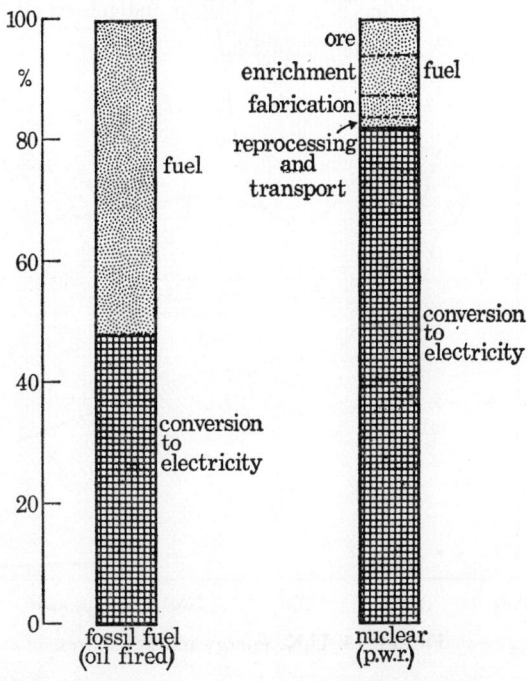

FIGURE 14. Cost structure of electricity production.

Figure 14 shows the relative delivered fuel costs and other costs of producing electricity from fossil fuel and nuclear fuel. The fossil fuel supplier plays a major role in both the cost and reliability of electricity production. With nuclear power the fuel supplier has a lesser role. Ore costs and quantities are small enough for stocks to be kept in terms of years rather than weeks of consumption. For practical purposes nuclear energy is indigenous, and at present electricity generation is the only route by which it can be brought into the nation's energy economy. Electricity producers, therefore, assume more of the role of primary energy producers. As time goes on this implies a broadening of their responsibilities and outlook – asking not only 'Are we supplying the demands of our customers reliably and economically?' but 'Are we making the right contribution to the energy resources of the country?'

Figure 15 shows a speculation about the market shares of secondary energy, in terms of useful energy delivered to consumers. The coal, oil and gas areas shown excludes their use for generating electricity. To 1972 it is factual. For the future the key assumptions are:

(1) Total end energy use growing at about 4% per year (corresponding to a lower figure for primary energy).

(2) Coal use, except for conversion, e.g. at power stations, falling to 20×10^6 tonnes per year by the end of the century.

(3) Gas levelling off at 20×10^{17} J per year.

(4) Electricity growing at about 5% per year.

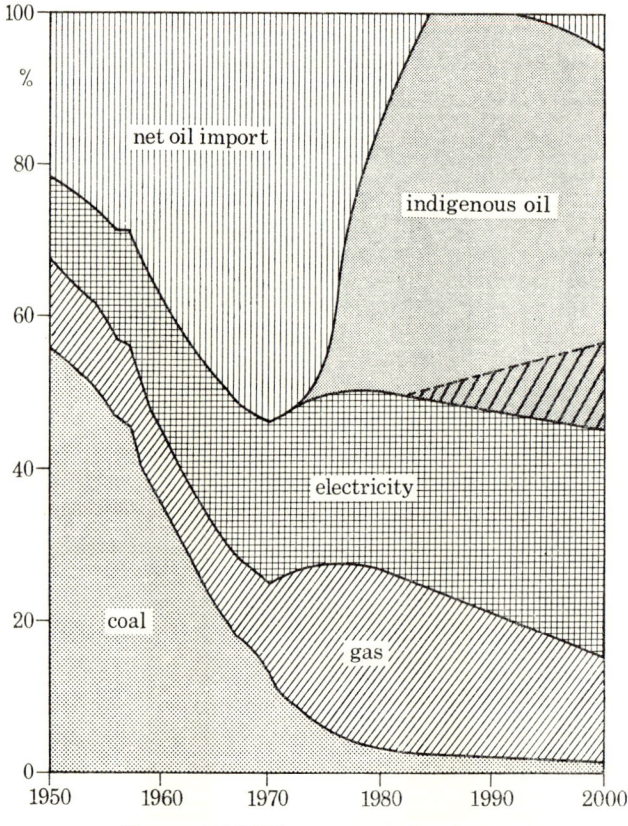

FIGURE 15. U.K. energy market shares.

With this projection the electricity share of the market would grow to 30% by the end of the century; and nuclear generation would be meeting 23% of the total end energy supply.

Oil consumption would still be high, even after allowing for falling use at power stations. The figure shows an optimistic view of indigenous oil production; it might fall well short.

The question arises: are the fossil fuel prospects so gloomy that faster expansion of nuclear power should be the aim?

By way of illustration, the cross-hatching shows electricity taking a larger share of the secondary energy market so that by the end of the century nuclear generation would be supplying 34% of the country's end energy consumption. This would mean boosting electricity growth to 7% per annum from the early 1980s. It would also mean substituting electricity for oil over something like 20% of the oil market. The lead time required would be long. The need for it hinges largely on the longer term prospects for oil from the continental shelf. Energy decisions of this kind call for some heroic forecasting. The views of the oil industry would be of the greatest interest.

Discussion

PROFESSOR E. EISNER (*Department of Applied Physics, University of Strathclyde, Glasgow*)

Use of rejected heat. May I start by thanking Mr Clark for a contribution of great interest and exceptional clarity, and particularly for slides that one could easily see and understand? I have two questions for him.

(1) In comparing capital investment costs and running costs, what interest rate is used?

(2) I have learned from Mr Clark that, by 1990, the Central Electricity Generating Board is proposing to reject about 200 GW of heat from its power stations. What is it proposing to use this heat for?

I was delighted to see the slide showing the flourishing oyster beds at Fawley, but disappointed to see that Mr Clark only regarded this as showing that waste heat from power stations did not need to be destructive. Why be so defensive about it? Why not regard the rejected heat as an asset whose use must form a part of the calculation of the use of existing power stations, and the design of future ones? And why not regard the oysters at Fawley as a demonstration of what some of us have been advocating for years?

Molluscs, and particularly the mussel, are among the most efficient known converters of sunlight into animal protein. Their efficiency of conversion depends strongly on water temperature, largely because of the rate of growth of phytoplankton. At present, the growing season in Britain is largely limited to about 3 months, but this would not be at all necessary if the mussels were grown in a fairly confined fall-out from a large power station. We spend some £200 million of foreign exchange a year on animal feedingstuffs, and I suggest that it is at least possible that we could easily save all of this using existing power stations.

I am sure that similarly valuable uses could be found for a very large part of this heat, which is at the moment wasted. In inland power stations, it seems to me that there is a prima facie case for considering horticulture in ground heated by the pass-out water. Any one who has experienced the poor quality and choice, and the high price, of lettuces and other horticultural products in Scotland will share my view that it will probably pay now, and even if it does not, it soon will with the rise in world food prices.

These, however, are only the sorts of things that could use heat from existing power stations, passing out at, say, 30 °C. If the principle of selling heat were an integral part of the design of new power stations, I am sure that the sale could profitably be at a price that would attract customers, especially with the rising price of fuel. Naturally, one would have to pay more per joule for warmer water. If such customers were sought as a matter of routine in the design of power stations, the enormous capital cost of the low-temperature end of generation could be saved, as well as the cost of getting rid of the heat.

What I am asking for is that the use of the low-temperature heat should be treated as part of the system being designed.

What work is the C.E.G.B. doing on these questions?

MR R. V. WATTS (*The Hatfield Polytechnic, Hertfordshire*)

Sir Eric Drake, in his paper, demonstrated that natural oil supplies were unlikely to be able to meet the demand from 1978 onwards. Oil has considerable value as a source of raw materials, as well as energy, it is also the most convenient source of energy for transport applications. These factors are likely to lead to available oil reserves, whether indigenous or not, acquiring considerable economic value and it was particularly disappointing to see the significant planned increase in the use of oil for electricity generation by 1981. As Mr Clark himself demonstrated, the electricity generating industry is well able to use alternative energy sources, in particular, it is probably the only major industry capable of making direct use of the large reserves of coal previously mentioned.

With regard to problems of energy storage associated with the increasing proportion of electricity from nuclear power stations, attention should be drawn to the considerable market for hydrogen implied by the proposals for the manufacture of s.n.g. and synthetic crude from coal. Thus, although, the 'hydrogen economy' may be too distant to consider very seriously, there may be scope for an indirect contribution by nuclear energy to the non-electricity market, by the production of hydrogen in off-peak periods for processing purposes.

SIR PETER KENT, F.R.S.

Mr Clark raised the question of whether it was reasonable to regard the present U.K. oil imports as replaced by the equivalent amount of indigenous oil from about 1977 onwards (his figure 15). In answer, Sir Peter Kent suggested that although the total quantities in the North Sea might well be correctly forecast (although including much oil as yet undiscovered), the physical conditions, the necessity to stretch engineering technology far beyond existing standards, and delays in availability of scarce highly specialized marine equipment would all operate strongly to produce slippage in off-take dates. The oil may not, in fact, be available as soon as was optimistically assumed.

Beyond this, it could conceivably be future policy that once the immediate energy gap has been passed, the remaining oil might be produced at a lower rate to make it available over an extra 10 or 20 years for specialized consumption.

There is, in consequence, good reason to believe that in Britain nuclear power will be required in greater quantity than shown on Mr Clark's minimal forecast, with an escalation of 7% per annum or rather more.

Phil. Trans. R. Soc. Lond. A. **276**, 571–586 (1974)
Printed in Great Britain

Operational experience in nuclear power stations

By T. Broom and R. S. Gow
Central Electricity Generating Board, London

From the first self-sustaining nuclear reaction to the present day represents a span of three decades: within that time large-scale generation of electrical power from nuclear energy has become acknowledged as economic, safe and environmentally acceptable. Within the U.K. 10% of electricity consumed is of nuclear origin. Some of the C.E.G.B. reactors have been in service for over 10 years.

The operating experience that has been gained shows how the original design concepts have been ultimately developed. Some of the difficulties encountered and the engineering solutions are presented. Operating experience feeds back to the design philosophy and safety requirements for future nuclear plant. In this way a foundation is provided for the further exploitation of what must become a major source of energy in the next decade.

1. The history of nuclear power in the U.K.

The equivalence of mass and energy which was established by Einstein at the beginning of this century was followed by the observation of energy release in radioactive materials and by element transmutation in the particle physics field. At that time the bombardment of nuclei by high-energy particles did not give any obvious lead as to how the energy released might be utilized. In 1938 it was established that fission of uranium was taking place in the natural state. The possibility of neutron release was realized and chain reaction appeared to be possible. The Maud Committee reported to the U.K. Government in 1940 its conclusion that it was possible to make a uranium bomb. The whole emphasis was on uranium-235 production, but it was soon recognized that plutonium could be used in weapons. The first chain reaction was set up by Fermi in the United States in 1942.

The post-war British effort was aimed at obtaining an independent nuclear deterrent and a working power programme. The U.S. approach to plutonium manufacture was to use water-cooled graphite-moderated reactors (Hanford) whose safety in U.K. conditions was questioned. Gas-cooled reactors seemed the best alternative, with aluminium-canned fuel. Although some empirical data on neutron absorption cross-sections were available they had to be checked and extended. Experimental reactors were therefore built at Harwell. This research yielded information that the data on the nuclear characteristics of magnesium were considerably in error and more favourable than had been reported so that its use as a canning material became possible. A series of magnesium alloys, generically known as 'magnox' was developed for fuel cans. Meanwhile, construction of the Windscale piles had taken place, using air-cooled aluminium-canned fuel. The Calder project, using magnox cans, was then initiated. During the whole of this period and up to the 1960s the lack of availability of enriched uranium dominated the scene so that no options were open to construct any reactors requiring enriched uranium such as light-water reactors. It was felt that the power programme would be based on the use of natural uranium fuel for a long time unless fast reactors could be quickly developed. To this end the Dounreay fast reactor was designed in the early 1950s. During the early period of Calder construction discussions took place between government and manufacturers and initially four consortia were set up to tender for the construction of nuclear power stations for the Electricity Supply Authority. The reactors could only be of the magnox type. As a result of the first round of tendering, the nuclear power stations at Berkeley and Bradwell were put on order.

2. The magnox programme

Berkeley and Bradwell went to power in late 1963 and since then six other magnox stations have been commissioned. The last set at Wylfa was commissioned in March 1973. All the magnox stations comprise two reactor units and associated turbo-generating equipment.

Design data for the C.E.G.B. magnox nuclear power stations are given in table 1 in chronological order of commissioning. Stations up to Sizewell utilized steel pressure vessels for reactor containment; Oldbury and Wylfa have steel-reinforced concrete pressure vessels. The number of coolant circuits has decreased but a minimum of four has been set. Turbo-generator capacity has also increased, although at Wylfa, with the much larger reactors, four generators of the proven size were installed.

During the magnox construction programme the aim was to reduce the cost per unit sent out at successive stations rather than to aim at operational advantage from replicated designs. Fuel rating, gas pressure and unit size were increased and the number of gas circuits reduced. Steam cycle efficiency was also raised under design conditions. This has meant that each station has incorporated some prototype design features giving rise to individual problems. Some examples are:

(1) Development work was required at most stations with on-load fuel handling equipment. Although fuelling chutes, which are introduced into the reactor only during the refuelling operation and are of fairly complex design, have performed satisfactorily with few exceptions, the more permanently installed pantograph and standpipe assembly components have required redesigned replacements at several stations. Fuelling machine availability has often been affected by poor performance of relays and electrical interlocking equipment, and marked improvements have been obtained during operation by substitution of small components with improved reliability.

(2) Although the testing of gas circulators during construction becomes prohibitively expensive as size increases so that rig tests must be confined to model tests on the larger sizes, their performance has been fairly good. At Hinkley Point 'A', part of a main gas circulator disintegrated during initial tests and all the circulators had to be strengthened and modified: performance since then has been good.

(3) Vibrations in gas circuits have been experienced but overcome by the provision of flow vanes at the elbow bends or by support improvement. Pressure-vessel insulation damage due to vibration at Wylfa involved temporary removal of a charge of fuel during the commissioning period; modifications to insulation mounting were carried out to prevent recurrence.

(4) Oil ingress via the gas circulator oil seals is serious because of its effects on the chemical reactivity of graphite and on the fuel-element heat transfer because of the formation of carbonaceous deposits. It has been a problem at several stations but has been largely overcome by further development of the rotating seals.

(5) Some developments have arisen from more demanding assumptions being introduced into economic and safety reappraisals, rather than by direct development of original designs. On-line delayed-neutron protection equipment is now installed at all stations with steel pressure-vessels to trip the reactor in the event of massive fuel-can failure.

TABLE 1. C.E.G.B. MAGNOX NUCLEAR POWER STATIONS OUTLINE DESIGN SPECIFICATIONS

basic data	Berkeley	Bradwell	Hinkley Point 'A'	Trawsfynydd	Dungeness 'A'	Sizewell	Oldbury	Wylfa
station s.o. capacity MW/reactor heat rating MW	275/558	300/531	500/971	500/860	550/840	579/948	600/892	1179/1875.5
mass of U per reactor/t	231.45	239	349	293	303.8	320.8	293	595.41
no. of fuel channels/no. of elements per channel	3265/13	2624/8	4500/8	3740/9	3932/7	3788/7	3320/8	6156/8
gas pressure (vessel inlet)/MPa	0.93	1.03	1.38	1.76	1.96	1.93	2.50	2.76
no. of boilers /gas circulators	8	6	6	6	4	4	4	4 sections/4
bulk gas outlet temp./°C	345	390	378	392	410	409.9	412	413.7
fuel element max. can temp./°C	432	440	437	435	452	453	450	451
fuel element max. rating/MW t^{-1}	4.1	3.98	4.37	4.97	4.58	4.74	5.1	5.03
total mass of machined graphite per reactor/t	1935	1811	2500	1972	2143	2200	2100	3735
type of gas circulator	induction motor, fluid coupling	induction motor, variable frequency	induction motor, variable frequency	induction motor, by-pass and throttle valves	back-pressure steam turbine	induction motor, vane control and gas by-pass	back-pressure steam turbine	induction motor, variable inlet guide vanes
h.p. steam/°C, at boiler s.v.	2.21/322	5.37/372	4.75/363	6.72/375	9.71/393	4.80/391	9.65/400 o.t.	4.83/396 o.t.
l.p. steam/°C, at boiler s.v.	0.54/322	1.47/372	1.35/349	2.16/365	4.06/395	1.94/390	4.86/393 o.t.	—
no. of main turbo-generators × capacity/MW	4 × 83	6 × 52	6 × 93.3	4 × 145	4 × 142.5	2 × 324.75	2 × 312	4 × 333.6
no. of aux. turbo-generators × capacity/MW	—	3 × 20.25	3 × 33		—	—	—	—
emergency generators type, no. × capacity/MW	diesel 4 × 0.6	diesel 3 × 0.450	diesel 5 × 0.780	diesel 4 × 1.2	diesel 4 × 1.64	diesel 4 × 0.747	g.t. 3 × 2.5	g.t. 3 × 3
contractors	T.N.P.G.	T.N.P.G.	A.P.P.	U.P.C.	T.N.P.G.	A.P.P.	T.N.P.G.	A.P.P.

o.t., once-through boiler.   g.t., gas turbine.

[167]

TABLE 2. C.E.G.B. MAGNOX NUCLEAR POWER STATIONS OPERATING RESULTS TO 31 MARCH 1973

station	end of commissioning, all turbines operational	net design electrical output MW	net maximum achieved electrical output MW	net accepted electrical output at 360 °C MW	days from end of commissioning to 31 Mar. 1973	total output up to 31 Mar. 1973 10^{15} J (10^9 kW h) †	cumulative load factor (%) against: design output	accepted output
Berkeley	Nov. 1962	276	302	276	3789	72.1 (20.02)	79.8	79.8
Bradwell	Dec. 1962	300	325	250	3789	74.1 (20.57)	75.4	90.5
Hinkley Point 'A'	May 1965	500	531	460	2870	77.1 (21.42)	62.2	67.6
Trawsfynydd	May 1965	500	508	390	2872	71.6 (19.89)	57.7	74.0
Dungeness 'A'	Dec. 1965	550	551	410	2648	88.4 (24.58)	70.3	94.4
Sizewell	Sept. 1966	580	496	420	2389	68.2 (18.96)	57.0	78.7
Oldbury	Sept. 1968	600	555	400	1643	45.3 (12.59)	53.2	79.8
Wylfa	Mar. 1973	1180	843	840	12	0.61 (0.17)	49.5	69.6

† Units supplied since commissioning of last set at station. In assessing the total benefits and costs of the magnox programme, rather than the operational experience, the units supplied prior to commissioning the last set should also be considered. This would reduce these load factors.

(6) Secondary shutdown facilities by means of the injection of an absorber in the form of boron steel balls are now available. Provision is also being made for the manually controlled injection of boron dust should permanent shutdown be required.

3. STATION PERFORMANCE SUMMARY

Table 2 summarizes the operating results. The operational time and the units generated have been measured from the time when the last set was commissioned with both reactors operational. The load factor is then the actual generation as a fraction of the maximum possible generation if running continuously under design conditions. As a consequence of unexpectedly rapid oxidation of some steel components, gas-temperature limits have been imposed which have caused derating at all stations, except Berkeley, since 1969. The net output which can be expected from the stations under these limits with all plant available and with reoptimized parameters is known as the accepted output capacity. Load factors given in these terms are also shown in table 2. The magnox stations have very low operating costs, once built, and are therefore in demand for generation throughout the year since the minimum system load demand exceeds their aggregated capacity. The station load factor figures given in table 2 are therefore almost synonymous with station availabilities – availability being defined as the fraction or percentage of the time during which the plant item concerned was available whether used or not.

TABLE 3. PERFORMANCE DURING PEAK LOADING WINTER PERIODS
(NOVEMBER–FEBRUARY 1962/3–1972/3)

station	net design electrical output/MW	winter electricity supplied (net) 10^{15} J (10^9 kW h)	winter days	cumulative load factor (%) against: design output	accepted output
Berkeley	276	26.3 (7.31)	1203	91.7	91.7
Bradwell	300	27.8 (7.71)	1203	89.0	106.8
Hinkley Point 'A'	500	26.5 (7.36)	962	63.7	69.3
Trawsfynydd	500	28.8 (8.02)	962	69.4	89.0
Dungeness 'A'	550	32.6 (9.06)	901	76.2	102.2
Sizewell	580	27.2 (7.57)	842	64.5	89.1
Oldbury	600	20.7 (5.76)	601	66.5	99.7

In the U.K., peak demands occur in the winter and the performance of the Board's plant is most critical at this time. The performance of the magnox stations during November to February inclusive is shown in table 3. The biennial consent to operate includes an assessment of the state of both reactors at a given station, assuming a further year of normal operation including 1 month of operating at an outlet gas temperature of 380 °C (20 °C higher than the normal gas outlet temperature). It is therefore possible to make use of this month during the winter period. This can result in the accepted output rising above 100%, as shown for Bradwell and Dungeness 'A'. The high load factors shown in table 3 confirm the confidence placed in the magnox stations by the C.E.G.B.

Operational availability is shown in figure 1. In this presentation it has been found convenient to show operational time, planned outage time, and forced outages separated into external and internal causes. External causes are those arising from plant outside the reactor–boiler complex and internal causes those within it. Figure 1 shows that with the single exception of

Hinkley Point 'A' in 1970–2, the forced outage total due to both internal and external causes is always less than the planned outage time. At Hinkley Point 'A' a major turbine failure led to prolonged outages (Kalderon 1972; Gray 1972).

The length of planned outages in the magnox reactors is principally determined by the time taken to carry out the statutory inspections required. Each station must shut down one of its reactors each year for boiler gas-side inspection, for which man-access is made. Water-side inspection is also required although this can be carried out on load on individual circuits.

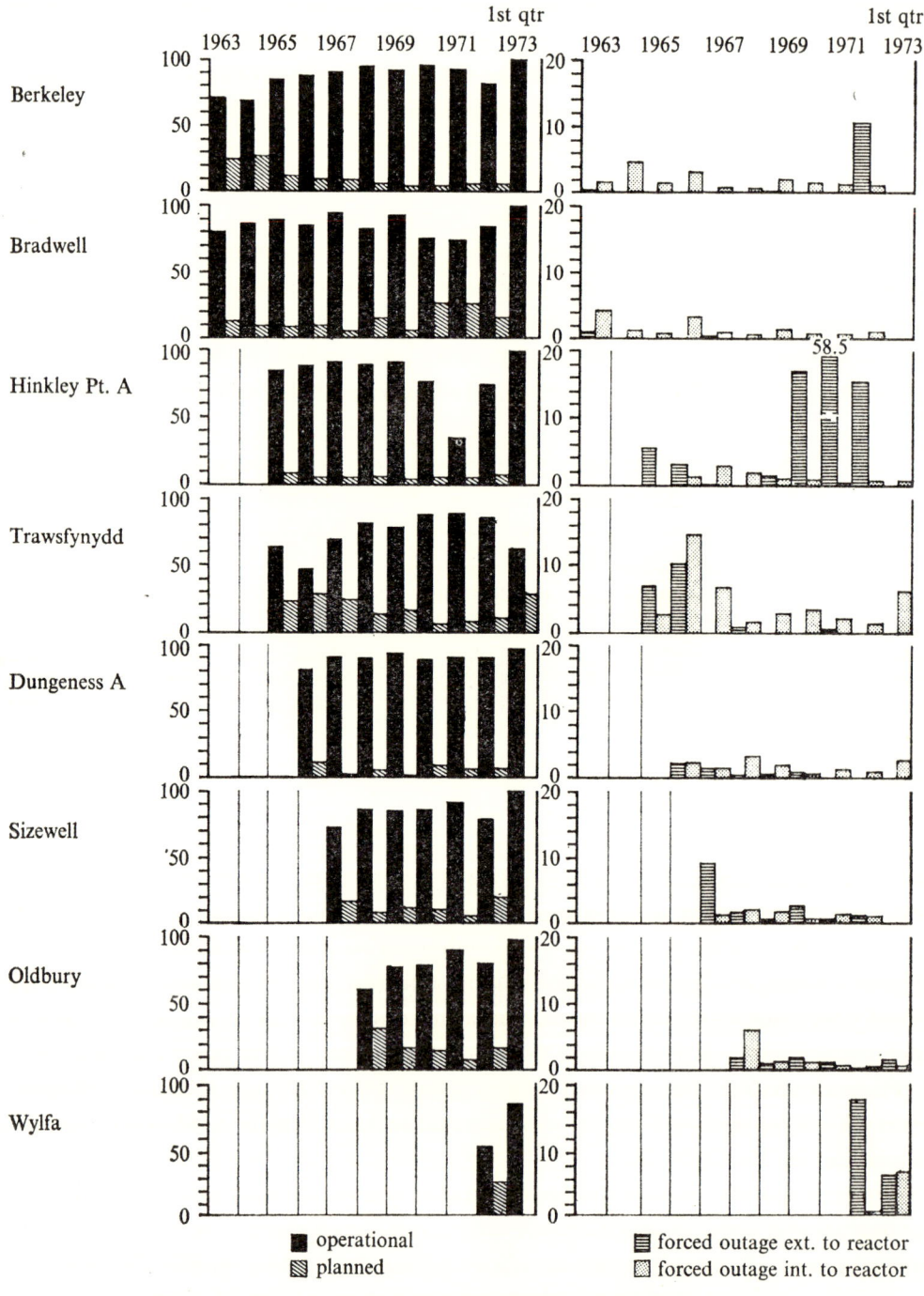

FIGURE 1. Percentage operational availability of C.E.G.B. reactors.

Reactor internal inspection is also a statutory requirement, carried out by a variety of techniques. The tendency is for planned outages to decrease in duration as operators improve performance through experience and forward planning: increased times may, however, occur as a result of increasingly onerous inspection requirements. The reduction of duration of planned outage times is generally the most profitable area for management concentration, and considerable efforts are directed to this end throughout the year.

Figure 1 also shows that the down time due to external forced outages generally exceeds that due to internal forced outages. These two components together with the planned down time add to the station operational availability figure to give 100%. The load factors which are shown in table 2 approach but cannot exceed the station operational availabilities. The causes of these outages have been further analysed in terms of plant items (Gow 1973). The figures speak for themselves and must be judged in comparison with the performance from other types of reactor or from other types of electrical generation. The high cumulative load factors and winter period figures indicate a satisfactory performance.

The running cost of nuclear generation in October 1973 was about one-third of the cost of generation at the best oil-fired and coal-fired stations. Appraisals of total costs per unit of electricity sent out depend on analyses of capital investments that were started up to sixteen years ago. Published figures (Hansard 1972, 1973) show that nuclear generation has been cheaper than at new oil and coal stations.

4. OPERATIONAL PROBLEMS ENCOUNTERED

No power-generation systems are without problems. Two of those encountered and dealt with in the magnox reactors are outlined.

Prior to the magnox programme, the technology of high-temperature gas cooling was investigated. Although large research programmes on material behaviour were started with the intention of covering the whole field, certain areas were not fully explored and this has led to problems in operation. The most important have been deformation of the fuel cans and mild-steel oxidation in hot carbon dioxide.

(a) Fin deformation

Reactor control is based on the measured heat transfer and pressure-drop performance of the fuel elements. Any fin deformation during service has serious implications since it may lead to a reduction in heat transfer and an increase in the pressure drop across the fuel element.

Some deformation has been produced by the stretching and buckling of the fins of high-temperature elements by an interaction between magnox and coolant gas (Harris 1972). This was a novel demonstration of creep produced by superficial oxidation stresses. In 1967 post-irradiation examination revealed significant fin deformation on Dungeness 'A' fuel elements at a channel average irradiation of only 10^{14} J/tonne (1200 MW d/t – megawatt days/tonne). Subsequently similar fin deformation was observed on fuel elements from other stations but in no case was the problem so severe as that at Dungeness. Fuel-element examination during discharge showed the axial and radial distribution of fin deformation in the reactor and hence the temperature-dependence of deformation.

Laboratory investigations defined the effects of gas composition and temperature and the effect of fin deformation on fuel-element heat transfer. All these results led to revised operating conditions. The proportion of hydrocarbons in the coolant was reduced by improvements in the efficiency of blower seals and the water content of the coolant was reduced by the use of dryers. At stations, such as Dungeness 'A', where the heat-transfer effects were significant, maximum fuel-element temperatures were reduced. Various design modifications were considered and a herringbone design of can, which is more resistant to fin deformation, was adopted for replacement fuel at Dungeness 'A'. However, before these design changes could be fully developed the effects of steel oxidation became apparent (Rutter 1972) and the temperature reductions which followed were sufficient to reduce the rate of fin deformation to a level which is not operationally significant.

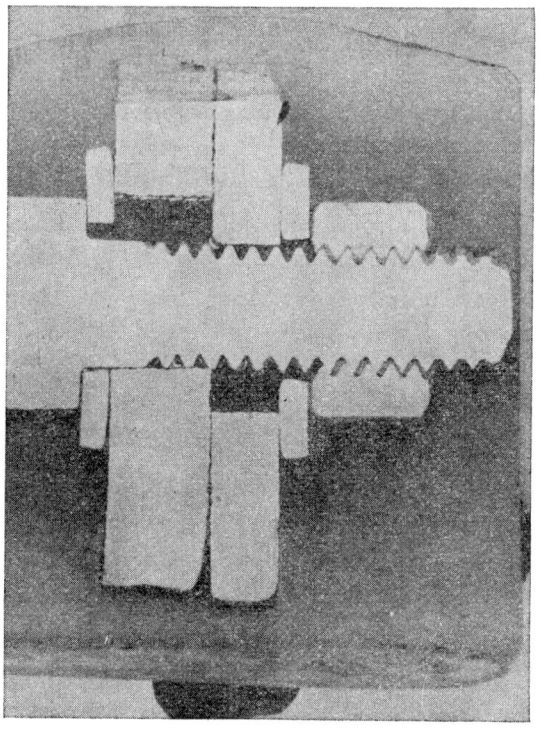

FIGURE 2.

(b) Mild-steel oxidation in hot carbon dioxide

The magnox reactors were designed for and originally operated at maximum bulk gas temperatures of up to 410 °C. It had been expected that oxidation of mild-steel components in carbon dioxide would be protective, with the rate decreasing with time as diffusion through the thickening oxide film became slower. In 1968 it was realized that 'breakaway oxidation' was occurring in some reactor components, with an enhanced and linear growth of oxide taking over from protective growth. If steel surfaces were originally covered with carbonaceous films then breakaway kinetics could be established at the outset. These observations led to decisions to limit maximum bulk gas temperatures to 360 °C to maintain an economic life for the reactors. Considerable research has led to better definitions of the major controlling parameters, temperature, wetness of the CO_2, and steel composition – and to the identification of a new oxidation mechanism (Rutter 1972; Gibbs 1973).

The most significant operational experience has been the straining and occasional failure of certain bolted assemblies, some of which include steels of low silicon content which are particularly susceptible to breakaway oxidation, some of which were coated with graphitic films during assembly, and some of which, because of the exigencies of constructional procedure in non-critical areas, include multiple washers and some misalinements. Growth of oxide between mating interfaces causes bolt extension (figure 2). Monitoring of the continued integrity of critical bolted joints is therefore essential, coupled with reliable forecasting of performance until the next scheduled reactor inspection. As there is no effect on oxidation of radiation, useful information can be obtained from boiler components. Means of inspection and sampling of reactor components are referred to in §5.

5. Reactor maintenance

(a) General considerations for vessel work

Conditions are met in nuclear plant which can create hazards different from those experienced in other branches of industry. Nuclear radiation is not detectable by any of the human senses, and working methods must therefore be controlled administratively with regard to instrument readings. Nuclear fuel continues to generate heat in a shut-down reactor because of fission-product decay heating. The carbon dioxide gas used in reactors is toxic and complete purging on shut-down may not be desirable or achievable.

All these circumstances show the necessity of breathing equipment for the man-access made to low-radiation zones in ducts and boilers associated with steel reactor vessels. Engineering requirements often dictate that access arrangements are difficult and working space is constricted. Provision for inspections should, wherever possible, be made at the design stage.

Concrete pressure-vessels, prestressed by tensioned steel ligaments, have been used for the later gas cooled stations. Their additional strength and accommodation of higher working pressures makes them suitable for urban siting. The reactors, boilers, circulatory and gas circuits are all contained within the pressure vessel as shown in figure 3. To gain access the carbon dioxide must be removed and replaced with air, and special air-cooled clothing has been developed because the temperatures cannot be easily reduced to atmospheric values with an integrated reactor boiler design. Under these conditions a man sweats approximately 0.5 kg water/h which is absorbed in the graphite. A graphite dry-out period of several days is required before start-up at the end of maintenance outages. Experience has been that circuits have remained clean and few problems have arisen with radiation dose or high levels of contamination within the boilers.

The cost of a shutdown of a magnox station in the C.E.G.B. system will vary between £20 000 and £60 000 per day according to its size, the time of year, and the availability of plant elsewhere. Downtime is therefore extremely expensive and all possible steps are taken to minimize it.

(b) Television inspection and photography

Reactor in-pile visual inspection equipment of several different types has been used, with emphasis on television inspection and photography rather than direct or mirrored viewing. Television viewing is suitable for general-purpose supervision because of the facility which it offers for immediate adjustment in position, focusing and zooming. It is also suitable for close-up work because of the ease of adjustment, and because in close-up work the definition may be adequate: for viewing a large area from a distance in detail conventional photography is used.

[173]

All stations have television equipment capable of being used to view the chargepans and reactor dome together with special television for inspecting the interior of channels.

For oxidation inspection, it is necessary to carry out very extensive photographic surveys at infrequent intervals, the surveys being planned in advance and occupying several days during reactor overhauls. In this way complete photographic surveys of reactor chargepans are carried out to identify any component movement or chargepan debris. The reactor top dome standpipe openings, pressure-vessel insulation and gas-duct openings have been similarly completely surveyed.

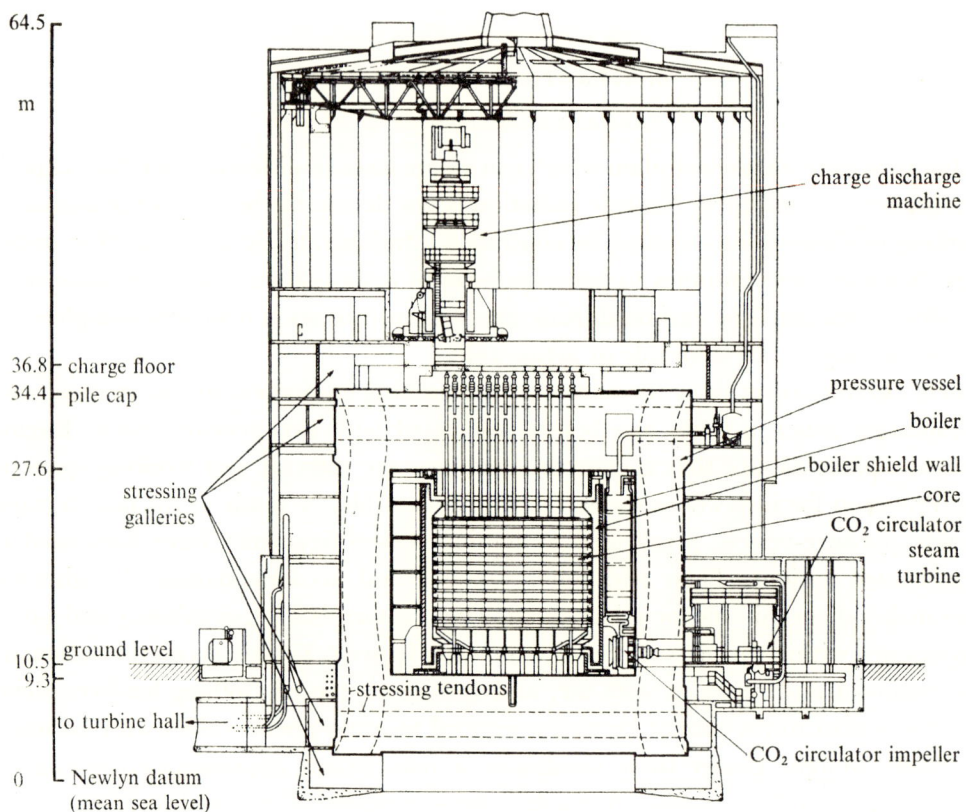

FIGURE 3. Oldbury: arrangement of reactor and prestressed-concrete pressure vessel.

(c) Oxidation inspection

When steel oxidation problems developed in magnox reactors it was necessary to increase the programmes of reactor inspection. Inspection is generally carried out in carbon dioxide with the reactors shut down, depressurized and cooled below about 70 °C. Attention is concentrated on those parts of the reactor which have been operated with carbon dioxide temperatures above 360 °C. Special manipulators have been developed for the handling of samples, tools and inspection equipment.

Ultrasonic testing has been successfully applied to bolts; impact and torque testing have also been used. Steel samples have been removed by grinding, trepanning and drilling. A pulsed laser technique is used for remote measurements of oxide thickness (Klewe *et al.* 1972).

(d) In-vessel engineering

There have been a number of reactor in-vessel component failures in this country and abroad of varying severity. Entries have been made to reactors containing large numbers of fuel elements. More particularly the development of remote-handling equipment and techniques have been accelerated and considerable experience has been gained. A problem arose at Trawsfynydd when a fuel chute could not be collapsed by the normal means. The chute was dismembered by remote tools and subsequently removed – an operation which took about 10 weeks to carry out. At Bradwell a steel sample basket was removed which had to be transported from the area at the top of the pressure-vessel down and out through one of the gas ducts. The most extensive work of this sort was the reinforcing of the top layer of the Bradwell core restraint. Twenty-four interlocking beams were lowered sequentially down standpipes and positioned around the top of the core to form an accurately mounted and tensioned polygonal garter. This work took about 4 weeks to complete and has now been done on each reactor. In work of this sort it is essential to manufacture a mock-up prior to initiating the actual work so that both the appropriate tools and the expertise of the staff can be developed.

(e) Safety-circuit and control-rod testing

The reactor is shut down in the event of accident by the insertion of control rods. These control rods are tripped in after the fault has been sensed by instruments which measure temperature, flux, pressure, and other relevant parameters. All reactor safety equipment must be engineered to a very high standard of reliability on a fail-safe basis. Redundancy is engineered into the equipment, usually in 2 out of 3 tripping logic. Maintenance of the reactor safety equipment must be to a very high standard and considerable effort is directed towards regular testing of this instrumentation.

Freedom of movement of control rods is of course essential and it is also necessary to be sure that their shut-down reactivity has not deteriorated and that they enter the core in the required time. Rod movement tests and rod timing tests are carried out on a routine basis. Rod reactivity may be inferred by other means but calibration tests are carried out from time to time if necessary, against xenon-135, which is a fission product whose absorption after shut-down is time-dependent. Analysis of control and instrument system maintenance performance is too extensive a subject to be covered here. The subject has been recently reviewed by Dixon & Jervis (1973).

6. NUCLEAR FUEL

The fuel elements used are of natural uranium, clad in a magnox can. The hottest fuel cans operate at peak temperatures of about 450 °C, with a peak centre uranium temperature of about 580 °C. Mean temperatures are considerably below these figures, the bulk gas outlet temperature being 360 °C.

(a) Fuel operational experience

The reactors each hold in total upwards of 20000 fuel elements. The residence time is 4–7 years, but is measured in 'burn-up' units of megawatt-days per tonne (MW d/t; 1MW d/t = 86.4 GJ/t). The fuel is changed on-load. Over one million fuel elements have been changed on-load, most of them at the end of normal life. About 0.1% of fuel elements have failed in service, mostly because of initial batch defects; and the present rate of failure is much less.

Fuel-element failure totals are shown in table 4 (Mummery & Hines 1973). Only 17 of these failures have resulted in any loss of output.

A considerable amount of development work has been necessary with fuelling machinery following first reactor start-up and changes in detail design of fuel elements themselves have required modifications during operation to refuelling machinery. The full capability of the fuelling machinery has not been realized at several magnox stations until several years of operation have elapsed, but despite these handicaps the plant load factor has been high.

TABLE 4. TOTALS OF FUEL ELEMENT FAILURES AT 30 SEPTEMBER 1973

	failure detection signal					
	start up and on-load charging	slow burst	rapid (1)	fast (2)	total	total fuel elements charged
Berkeley	20	129	17	3	169	297 206
Bradwell	14	89	19	6	128	122 156
Hinkley Point 'A'	65	84	12	3	164	180 544
Trawsfynydd	10	52	6	5	73	154 845
Dungeness 'A'	22	317	11	2	352	144 298
Sizewell	12	503	2	1	518	131 593
Oldbury	3	4	0	0	7	105 388
Wylfa	0	3	1	0	4	109 664
totals	146	1181	68	20	1415	1 245 694

NOTES: (1) Signal doubles within 8 h. (2) Signal reaches saturation within 1 h.

(b) Fuel development

Twelve of the 16 C.E.G.B. reactors have now reached the end of the lifetime of the first fuel charge and equilibrium cycles have been established. The performance of the fuel has been assessed by a programme of fuel-element monitoring carried out jointly by the C.E.G.B., U.K.A.E.A. and B.N.F.L. The principal purpose of monitoring has been to underwrite the safe operation of the fuel but an additional objective has been to acquire information to enable improvements in the efficiency of fuel utilization to be made. At present some 7000 elements have been examined out of a total of about 1 million which have been loaded into the reactors. These examinations cover 'burn-ups' ranging to about 5×10^{14} J/t (5700 MW d/t) channel average irradiation.

As a result of this work the irradiation target for fuel has been progressively raised from 1.47×10^{14} J/t (1700 MW d/t), through 2.6×10^{14} (3000) and 3.1×10^{14} (3600) to 3.45×10^{14} J/t (4000 MW d/t). The principal problems encountered have been uranium-bar torsion, fuel-element bowing, and uranium swelling, the latter being caused by fission product gas build-up. These occur generally late on in the fuel life and provide an ultimate limit to operation. There were some unexpected problems with magnox cans arising early in fuel life. The most important of these were the occurrence of numerous can wall failures in certain batches of fuel elements and the fin-deformation problem already mentioned.

It appears that as the causes of fuel-element failures are understood and improvements in manufacturing and operating techniques implemented, the existing very low rate of failure will be maintained or even decreased further. The deformation of the fuel-element heat-transfer surface due to fin waving was unexpected, but solutions to this problem are available should a return be made to high-temperature operation. Neither torsion nor bowing of the fuel bar are likely to lead to difficulties and it must be inferred that the most probable life-limiting feature is the onset of breakaway swelling at very high discharge irradiation.

The ultimate objective of a fuel-element development programme is that the fuel is not discharged until it has reached its reactivity limit – that is, the point at which criticality can no longer be maintained. The fuel in the outer region of the core is already very close to this limit, but is capable of further burn-up in the central region. The axial or radial shuffling of fuel is being considered further to exploit the fuel utilization. A fuel element irradiation increase of about 50% is theoretically achievable. It is of course necessary to balance the economic advantages of more efficient fuel utilization against the possible penalties associated with the increased risk of reactor outrage if fuel were double-handled and its residence times further increased. Experiments on shuffling and on other dual residence schemes are being carried out.

(c) Cooling ponds

One of the major areas of difficulty has been in dealing with irradiated fuel discharge within and in the environment of the reactor cooling pond. The irradiated fuel resides in the pond for a period of about 3 months and is then returned to the fuel manufacturers (B.N.F.L.) for decanning and chemical processing. Radiological control is achieved by the application of appropriate safety rules within I.C.R.P. requirements. Radiation hazards arise from the fuel elements themselves and from any activity which may be present in the pond water or on the surface. Provided that the fuel residence in the cooling pond is not excessive and that failed fuel elements removed from the reactor are bottled there is small probability of activity building up in the pond water. However, practical difficulties at times preclude these objectives being achieved and damaged fuel elements may allow fission products to escape into the pond water. Clean-up plant is provided, consisting of filtration equipment and ion-exchange units.

7. Environment

(a) Effluents

The Board have a statutory duty to have regard to the effects on the environment in discharging any of their functions. Nuclear legislation requires that radioactive effluents, gaseous and liquid, arising from nuclear power stations are rigidly controlled to within authorized discharge limits. The principal criteria are:

(1) No member of the public shall receive more than one-tenth of the I.C.R.P. level for occupational workers.

(2) The whole population of the country shall not receive an average dose of more than 1 rad (10^{-2} J/kg) per person in 30 years.

(b) Activity arising

Reactor operation gives rise to solid, liquid and gaseous forms of activity. The storage and treatment of solid and liquid activity are determined by whether the activity is in the form of radiation or contamination. Dry storage of irradiated redundant reactor components in concrete vaults on-site is used. Contaminated solids can also be stored on-site in suitably constructed active waste areas. Activity occurs on filters and resins used for ion-exchange treatment. Resin regeneration gives rise to active sludges, and waste dumps are used on-site for this storage.

A considerable volume of low-activity waste accumulates in the form of packing paper, cleaning rags, overalls, etc.: some of this may be combustible. Reduction of volume of a considerable amount of this low-activity solid waste is possible by incineration, the greater part of the activity being retained in the ash. This method is used under careful monitoring at the stations for this purpose. Low-activity oil accumulates from rotating plant used in reactor equipment and is disposed of by combustion.

Transport of highly active materials is thus reduced to the carriage of spent fuel and to small quantities of fuel and other items which are sent to the laboratories for examination. Even so, the policy has been maintained of bringing laboratory facilities to the site wherever possible when large-scale special examination of active components is necessary.

TABLE 5. ACTIVITY DISCHARGED IN 1972

| station | tritium liquids (Ci) | | non-tritium liquids (Ci) | | active aerosols (mCi) | |
	limit	discharged	limit	discharged	discharge	dilution factor at 0.5 km
Berkeley	1500	44.2	200	23.3	5.7	40
Bradwell	1500	251	200	119	3	40
			(5 ^{65}Zn)	(0.03)		
Hinkley Point 'A'	2000	38.6	200	147	43.4	40
Trawsfynydd	2000	46.0	40	31.4	33.9	15
Dungeness 'A'	2000	28.9	200	29	94.1	40
Sizewell	3000	53.2	200	14.6	79	35
Oldbury	2000	15.0	100	5.2	32	35
Wylfa	4000	82.7	65	0.3	3.2	40

Analysis of milk samples for iodine-131, caesium-137 and strontium-90, and dose measurements made outside the station boundaries, have demonstrated that the operation of the stations is causing no detectable changes to the environment. These monitoring surveys are presented to each nuclear station's local Liaison Committee with the object of keeping local residents fully informed. Activities discharged from the magnox stations in a typical year, 1972, are shown in table 5. The quantities of tritium released are very small, particularly in comparison with some water reactors (Richardson 1973)and the argon-41 activity in the shield cooling air decays rapidly. Fission product activity arising from failed fuel elements is minimal as these are discharged on-load as soon as failures occur and it has not so far been possible to correlate the observed district measurements on fission product activity with anything other than bomb tests. The zinc-65 limit at Bradwell has been set in view of the proximity of the beds of the local oyster industry. The aerosol figures are obtained by sampling through a filter paper: additional small quantities of sulphur-35 and tritium vapour were released. Gaseous argon-41 figures are not shown as this is not measured regularly: releases are of the order of 10^5 Ci per year per station, an insignificant figure for this isotope which has a short half-life.

(c) Radiation dosages to the public and to workers

Since Berkeley and Bradwell became operational in 1962 there has been no radiation exposure to individual members of the public exceeding the I.C.R.P. limits.

More than 5000 people are employed as radiation workers at nuclear power stations and particular importance is attached to radiological protection and monitoring. Radiation exposure is monitored by film badge for normal duties in normal radiation areas. These areas are such that full working residence in them could not result in acceptable dose rates being exceeded. When access is required to more active areas with higher radiation, individual dose meters are carried which give an immediate reading of the dose as it is integrated. When work is required in contaminated areas, precautions are necessary against ingestion through the body orifices. Precautions range from protective clothing to combat surface contamination to full enclosure with breathing equipment to combat air contamination. An annual medical test of classified workers is carried out, including blood tests.

8. Nuclear safety

Within the C.E.G.B. the Chief Nuclear Health and Safety Officer has direct responsibility to the Board, for the assessment of the safety of plant and the provision of advice on safety matters. His inspectors and technical staff stand apart from operational staff and constitute an independent line of surveillance.

The Chief Inspector of the Nuclear Installations Inspectorate (N.I.I.) advises the Minister and assesses proposals for new reactors and examines details of design and construction.

A site licence is necessary before construction can begin and this is issued by the N.I.I. after examination of comprehensive safety studies. The N.I.I. monitors the operation of stations through its own visiting inspectors and through receiving the formal proceedings of the Safety Committee which is appointed for each station.

The regulation of nuclear safety in England and Wales is by the Nuclear Installations Acts of 1965 and 1969 administered by the Secretary of State for Trade and Industry. C.E.G.B. sites are licensed by the Minister, who may attach special conditions relating to safety. The C.E.G.B. is entirely responsible for ensuring safe operation in respect of its own staff and of the public around the site. It must of course conform with the legislation and conditions imposed by the Minister and the immediate responsibility for this in each station lies with the Station Superintendents.

The Nuclear Safety Committee for each station is comprised of senior members of the staffs of the C.E.G.B., U.K.A.E.A. and B.N.F.L. Formal reports concerning safety matters are required and the approval of the Committee has to be obtained in circumstances where certain variations in operating procedures are proposed.

Outside the formal arrangements for day-by-day safety matters, the Minister is advised by a Nuclear Safety Advisory Committee whose members embrace all aspects of British nuclear experience. At individual station level, nearby communities and local authorities are kept informed of all relevant aspects of station activities through local liaison committees so that by this means a good understanding of safety issues can be achieved.

9. Conclusions

The whole of this paper has been based on C.E.G.B. experience. Against a world background, the cumulative total of nuclear electrical energy generated in the U.K. is 2.5×10^{18} J (694×10^9 kW h), about 37% of the world total.

During the next few years this percentage share will fall as other countries increase their nuclear capacity. In this country the advanced gas cooled reactors (a.g.r.) will be coming on line, the first of these in 1974.

The future of nuclear power generation lies with the breeder reactor, which offers the prospect of increasing the fission yield of natural uranium by two orders of magnitude.

Looking to the 1980s the C.E.G.B. is concerned that the next reactors shall be replications of proven designs. Important operational requirements are ease of inspection, access and maintenance, so that any necessary repairs can be carried out.

At the same time we need to accumulate operational experience of reactor types with future potential in demonstration sizes but would not expect to develop these entirely out of our own resources.

Society may have confidence that the nuclear power programme in this country will be soundly based on considerable operating experience and that the safety of staff and public will be assured.

Acknowledgement is made to the C.E.G.B. for permission to publish and to the authors' colleagues for assistance in the provision of information.

REFERENCES (Broom & Gow)

Dixon, F. L. & Jervis, M. W. 1973 *Int. Atomic Energy Agency* (paper SM-168/A-2).
Gibbs, G. B. 1973 *Oxidation of Metals J.* 7, 173–184.
Gow, R. S. 1973 *Int. Atomic Energy Agency* (paper SM-178/18).
Gray, J. L. 1972 *Proc. Inst. Mech. Eng.* (paper 32).
Hansard 1972 Question 54/72.
Hansard 1973 Question 35/73.
Harris, J. E. *et al.* 1972 *Int. Atomic Energy Agency* (paper 167).
Kalderon, D. 1972 *Proc. Inst. Mech. Eng.* (paper 31).
Klewe, R. C. *et al.* 1972 *J. Phys E. Scient. Instrum.* 5, 203–205.
Mummery, G. B. & Hines, G. M. 1973 *Br. Nucl. Energy Soc. Conf.* (to be published).
Richardson, J. A. 1973 *Am. Soc. Mech. Eng.* (paper 72/WA/NE-4).
Rutter, R. L. 1972 *Int. Atomic Energy Agency* A Conf. 49 P/468.

Discussion

Mr M. F. Osmaston (*The White Cottage, Sendmarsh, Ripley, Woking, Surrey GU23 6JT*)

Dr Broom has shown that, recently, the annual total electricity generated by the C.E.G.B.'s nuclear power stations has attained a very high percentage of the 'accepted' output capacity of these stations. One suspects that this is not so much a measure of the reliability of nuclear-power technology as it is a credit to the C.E.G.B.'s judgement of what is safe when agreeing station acceptance ratings. To set this in proper perspective, perhaps Dr Broom, in the published form of his paper, would give the accepted percentage derating relative to the originally ordered electrical output capacity for these stations, both overall and (if trends exist) as functions of station size and date of ordering. A comparable figure for fossil-fuel stations over the same period would also be of interest.

T. Broom

The information sought can be derived from table 2. The net total derating is 23 %. In respect of those fossil fuelled stations which have had their last turbogenerators commissioned during the period November 1962–March 1973 their present total net electrical output is 23 600 MW compared with a net design output of 24 950 MW, a reduction of 5 %. Improvements to these ratings are still being acheived.

Phil. Trans. R. Soc. Lond. A. **276**, 587–601 (1974)

Printed in Great Britain

Future trends in nuclear power generation

BY SIR JOHN HILL

United Kingdom Atomic Energy Authority, London

In the 20 years since the Calder Hall reactors were ordered, the U.K. has accumulated wide experience of building and operating nuclear power stations.

Early stations proved expensive because of technological novelty and infrequent orders, but the economics of nuclear power stations where regular orders can be assured are increasingly favourable. Other factors do not provide fundamental limitations to nuclear power growth.

Trends in fossil-fuel prices suggest that most utilities will look mainly to nuclear plant to meet their electricity requirements. The substantial savings of fossil fuel already achieved will thus grow rapidly on a world-wide basis. Though it would take quite unexpected shifts in relative economics for nuclear stations completely to supplant conventional stations, particularly for peak demand situations, a high nuclear share of new capacity may begin to throw some strain on uranium reserves in the 1980s.

The fast reactor, prototypes of which after long and careful development are commissioning in France, Russia and the U.K., can provide a huge increase obtainable from uranium resources, pending the successful introduction in the long term of fusion reactors.

Twenty years ago construction started of the nuclear reactors at Calder Hall and in October 1956 the first reactors there delivered electricity from a full-scale nuclear power station into a national distribution system for the first time in the world. Since then nuclear power has spread and grown in importance all over the world, particularly in the United States and Western Europe. Moreover, it is now clear that this growth will continue, at an accelerated rate.

In the European Community there is currently nearly 11 000 MW (e)† of installed nuclear capacity, 22 000 MW (e) under construction, and a further 11 000 MW (e) on order and at the planning stage. Estimates of future capacity vary, but some Community officials have called for a total capacity of 200 000 MW (e) by 1985 (E.E.C. 1973). Such an expansion, though justified by the energy trends already discernible, is probably not feasible, but there can be no doubt that a period of substantial expansion is beginning.

The United Kingdom is well placed to participate in this expansion. Since the inauguration of her nuclear power programme, this country has accumulated wide experience in the construction and operation of nuclear power stations. Following the construction of the four Calder Hall reactors and a similar set of four reactors at Chapelcross (all of which are currently operated by British Nuclear Fuels Limited with an output considerably in excess of the design figure) a series of nine commercial magnox stations were built, and have been operated by the Electricity Boards. Two further magnox reactor power stations were sold abroad, one to Italy and the other to Japan.

All these magnox reactors were of the thermal type. In a thermal nuclear reactor a moderator surrounding the fuel reduces the kinetic energy of the neutrons, increasing their chance of capture by a fissile uranium-235 isotope. In the resultant atomic fission heat is produced, and further neutrons are released which maintain the chain reaction within the reactor fuel.

These early stations have in general proved to be successful technically. Although other

† The power measured is the electrical output.

countries have in recent years been developing their nuclear power programmes at a far greater pace, this country has still generated more electricity from nuclear reactors than any other country (see, for example, Nicholson 1973), and this has been achieved almost entirely by the magnox reactors. The first Calder Hall reactor has now run without interruption, except for routine maintenance, for over 17 years, with a cumulative load factor of 88%. Several of the commercially operated magnox reactors, as indicated in Dr Broom's paper, have achieved impressively high cumulative load factors, together with high availabilities.

The magnox stations were succeeded by the advanced gas-cooled reactors (a.g.r.), which have many features in common with them but use slightly enriched fuel in stainless-steel cans, which permits operation at higher temperatures. A small experimental reactor of this type has been operating successfully at Windscale for over ten years, with a cumulative availability of 84% and a cumulative load factor of 71%. Five commercial stations will come into operation over the next few years. There have been problems and delays in their construction, but in many cases these have not been specific to the reactor concept. Delays have been experienced all over the world in the construction of generating plant, both conventional and nuclear, mainly stemming from the rapidly increasing size of the plant units. For example, while the first U.K. reactors had capacities of only 35 MW each, the a.g.r. have capacities of 600 MW or more each, and in America and elsewhere individual units of up to 1200 MW are being constructed. Similar increases have taken place in the size of associated plant such as turbines and alternators, and in that of conventional power units.

TABLE 1. FORECAST GROWTH OF NUCLEAR POWER IN THE PRINCIPAL WESTERN EUROPEAN COUNTRIES – GW (e) (FROM O.E.C.D 1973)

	1970	1975	1980	1985	1990
Belgium	—	1.7	3.0	5.5	10.0
France	1.5	3.8	13.4	32.5	67.0
Germany (W.)	0.8	4.9	19.0	38.0	75.0
Italy	0.6	1.4	6.0	18.0	44.0
Spain	0.1	1.1	8.0	12.0	24.0
Sweden	—	3.2	8.6	13.0	24.0
Switzerland	0.4	1.0	4.0	8.0	16.0
United Kingdom	3.4	8.8	13.8	35.0	75.0

The constructional problems with the a.g.r. are now being overcome. The good record of the magnox stations and the continuing encouraging performance of the Windscale reactor justify confidence that the a.g.r. stations will prove to be very valuable components of the country's generating system.

In the rest of Western Europe the nuclear reactors have predominantly been of the water-reactor type, although France for some years developed a gas-cooled graphite-moderated line of reactor development. The reactors have, in general, been built by national firms utilizing American technology under licence, although in some cases they have been constructed by an American firm directly or through a European subsidiary. While slower to start constructing nuclear power stations than the U.K., several nations in Western Europe, particularly France and West Germany, have since the late 1960s been constructing nuclear plant at a faster rate. The forecast programmes for these countries, and others such as Spain, Sweden, Italy and Switzerland, are ambitious, as table 1 demonstrates.

The early nuclear power stations proved to be relatively expensive for a number of reasons. The novelty of the technology, the special engineering machinery required, and the fine tolerances demanded, were all reflected in the high capital costs of nuclear stations. Capital costs have also been pushed up by sporadic ordering patterns, the fluctuating size of total programmes, and more recently by difficult problems associated with scaling up (by factors of up to 20 in some cases).

These differentially high capital costs have been emphasized by the steep increase since the mid-50s in the test discount rate used in the U.K. public sector. The 1955 White Paper on Nuclear Power was based on a public sector rate of 5%; Generating Boards are now required to use 10% in real terms (i.e. with constant money values).

In the future, given a steady and regular programme of orders, the economics of nuclear power stations will become increasingly favourable. Nuclear capital costs should fall in constant money terms. The use of bigger-sized plant, associated together in groups of, say, four units, should enable the advantages of economies of scale to be realized, and also enable savings through 'doubling-up' in the use of facilities and staffing. With a sufficient flow of orders there can be savings gained through replication. As more nuclear stations of each type are built, the high costs associated with the launching of a new reactor will be eliminated, and the costs of further reactors of that type will fall as more is learnt of their construction and operation.

This process of reduction in nuclear capital costs will be much more pronounced than any that can be expected in the capital costs of fossil-fuelled plant. This is largely because the technology of fossil-fuelled stations is older and most of the improvements possible have already been made. As a result the difference between the capital costs of nuclear stations and the capital costs of fossil-fuelled stations is expected to diminish, although nuclear stations will probably remain rather more expensive.

A second factor which should keep the costs of nuclear power down for this country is the existence now of an efficient British nuclear industrial organization to construct commercial nuclear stations, provide fuel, improve the performance of existing types of station, and develop new systems for the future.

On the construction side, it at one time seemed appropriate for there to be five nuclear consortia competing for orders to build magnox stations both in this country and overseas. But, particularly with the increasing size of individual reactors, domestic orders were insufficient to maintain so many consortia, and none of the consortia was able to call on sufficient resources to compete overseas with companies which either had the advantages of tendering in their own countries, or had the huge resources of the American firms. The number of consortia reduced progressively from five to four, then to three and finally to two, the last reorganization taking place in 1968/9, when British Nuclear Design and Construction Ltd and the Nuclear Power Group Ltd were formed. But even at the time many of us would have preferred a single-company organization, and there has been a growing realization that further rationalization of the industry was indeed necessary. The government has recently presided over a restructuring of the industry into a single reactor construction company, which is designed both to meet the domestic demand for nuclear stations and to compete and, equally important, collaborate, effectively overseas. Such collaboration is highly desirable, especially between the countries of Western Europe.

The opportunities for fruitful international collaboration and the advantages of a single strong organization have been demonstrated by British Nuclear Fuels Ltd, which was set up in

[183]

1971. The main activities of this company are reactor fuel element design and manufacture, irradiated fuel reprocessing, UF_6 conversion and uranium enrichment. In addition to supplying nuclear fuel and reprocessing services to the home generating boards, B.N.F.L. has secured a good deal of export business, particularly in the fields of UF_6 conversion and reprocessing. For example, reprocessing contracts have been obtained in Canada, Germany, Italy, Japan, Spain, Sweden and Switzerland.

The performance of the fuel elements has been exceptionally good and any delays in reactor operation have certainly not been attributable to late deliveries of fuel.

Two areas which particularly illustrate the benefits of international cooperation are those of uranium enrichment and fuel reprocessing. Both involve considerable capital expenditure and it is essential to ensure that installed capacity matches the available market. The development of nuclear power requires an assured supply of uranium enriched in the 235 isotope. The bulk of the western world's supply of enrichment at present comes from the gaseous diffusion plants in the U.S.A., but these do not have sufficient capacity to meet the firm demands foreseen in the 1980s. There is the further point as to whether it is politically acceptable for Europe to rely on a monopoly supplier in the U.S.A. The diffusion process is one that requires a very large plant before it becomes economically viable, and it consumes very large amounts of energy. Following independent studies of an alternative method of enrichment using the gas centrifuge which were carried out in Britain, Germany and Holland, the three Governments signed in 1970 a treaty concerned with the exploitation of this process. Two international companies were set up: Centec, with headquarters in Germany, for the development and manufacture of enrichment plants, and Urenco, whose headquarters are in this country, for the operation of enrichment plants and the marketing of enrichment. B.N.F.L. is the British shareholder in both these companies, which have recently been brought under a united senior management.

In the equally important field of fuel reprocessing, the existing plants in Britain and France have sufficient capacity for the immediate European needs with a margin for overseas business. In the not too distant future, however, additional plant capacity will be required and, so that this may be developed in an orderly way, an international company – United Reprocessors – with British, French and German partners has been set up. B.N.F.L. is also the British partner in this company. Transport of irradiated fuel from the reactor sites to the reprocessing plants is another area in which B.N.F.L. is cooperating with French and German partners.

All these collaborative arrangements were facilitated by the existence of a single company concerned with all aspects of the nuclear fuel cycle.

The Atomic Energy Authority provide the third component of the nuclear industry – the research and development organisation. The Authority work on the improvement of the performance of existing reactor designs and advise on their safe and efficient operation. For the immediate future, development work is devoted to new systems such as the steam-generating heavy water reactor (s.g.h.w.r.) and the high-temperature reactor (h.t.r.), the latter partly on a collaborative basis in the O.E.C.D. Dragon project. The largest component of the Authority's programme is, and has been for some years, the work on the fast breeder reactor, which, as I shall explain later, holds the key to a balanced nuclear system for the foreseeable future.

The single reactor construction company, the single nuclear fuel company and the single research and development organization are all closely interlinked to provide a potentially very strong nuclear industry. Shortly it will have the task of designing, constructing and fuelling the next round of nuclear power stations. It is not yet clear what type of reactor will be chosen for this round. There are four main contenders, namely the a.g.r. (the type now under construction), the s.g.h.w.r. and the h.t.r. (both the subject of development work by the Authority), and the light-water reactor, largely developed in the United States. The balance of choice between them is fine; the small differences in generating costs means that the options will remain open even if engineering or safety problems arise for a particular line of development. Once the choice is made, and a sustained programme of construction started, the industry's position will be firmly established.

Certainly only such a programme, affording scope for replication and a fair degree of continuity, will give the industry the opportunity to provide the experience and quality control essential for timely completion and close control of costs. Moreover, while such a programme should include exports of reactors as a major objective, the initial impetus can only come from the domestic market. The history of nuclear marketing both here and overseas demonstrates unequivocally that only a secure and adequate home demand can provide a base strong enough for marketing overseas and for sound international arrangements.

But perhaps the most important factor favouring the development of nuclear power in the next 15 years is the status of alternative sources of energy. Prices of these alternative sources of energy, particularly oil, have risen rapidly recently and seem likely to continue to rise for the foreseeable future. At the same time, several of these alternative sources have a discernible exhaustion point which in some cases appears to be relatively near. Both in the short and the medium term this is likely to bring about limits in supplies by the producing countries which will seek to conserve part of their resources for the future. Again this will be particularly true for oil.

Even with North Sea oil, the European Economic Community (including the United Kingdom) could depend on imported oil for some 50% of its energy needs in 1985, according to an official estimate made earlier this year, and the price of North Sea oil will almost certainly be determined by world prices.

Similarly, with the other principal conventional fuels used in Western Europe – hydro-electricity, natural gas and coal. Although there is some scope for expansion of *hydro-electricity*, this is insufficient to meet increases in energy demand. Indigenous resources of *natural gas*, even with recent discoveries, are limited and restricted to premium uses. Supplies of liquefied natural gas and synthetic natural gas are likely to be obtainable only at prices high enough similarly to restrict their use. *Coal* mining must remain essentially a labour-intensive industry despite increasing mechanization. If the men employed are to maintain a standard of living compatible with the task they are being asked to undertake, the labour costs of the industry will rise, which will be reflected in the cost of coal. Therefore, even if the price of power-station coal relative to oil becomes more favourable, that price will nevertheless rise. This is especially true for the non-British Western European coal industries, which face particularly severe problems of recruitment, rising costs and exhaustion of the resources.

In the longer-term there is a general problem of demand outstripping supply for all these fuels. To gauge the extent and urgency of this problem the likely future demand for energy must be set against the resources available to meet that demand.

By A.D. 2000 and assuming only modest rates of growth there could well be a short-fall of the order of 40% of world energy requirements. It is unreasonable to expect acceleration of production from traditional sources plus the exploitation of non-traditional sources (such as oil shales and tar sands) to fill more than half of this gap, and even this much implies the acceptance of real price increases in traditional fuels by factors of 2 and 3 (Brookes 1973).

The remaining requirement, of the order of 20% of total world energy needs, must for all practical purposes be met by nuclear and hydro-electricity. Because of the limitations on further development of hydro-electricity, some 90% of this requirement has to be met by nuclear power. On this basis a world installed nuclear capacity of some 5000 GW (e) would be required in the year 2000.

Western Europe's share of this capacity would be some 800–900 GW (e) (compared with 12.3 GW (e) in 1972). This is a daunting increase, but should not be impossible. And failure to meet it would severely strain fuel resources, with consequent setbacks for national economic growth-rates. If the target is to be attained it is necessary that a good start on the process of nuclear power expansion be made as quickly as possible.

Power stations installed now and fuelled by traditional fossil fuels commit the country to a significant long-term demand for an increasingly expensive and economically unfavourable energy source. A substantial delay in starting the build-up of nuclear capacity may lead to difficulties later in providing the necessary economic resources for the delayed expansion, which would come at a time when increased fossil-fuel bills were already straining national economies. The need for many users of energy to adapt themselves in the next decades to electricity instead of other fuels will in any case be a call on resources. Certainly, there will be an interim period before the large-scale installation of nuclear plant can cushion the effects of fossil-fuel price rises on economic activity. But if the interim period was prolonged the electricity production and many other industries would have no alternative but to face substantial fuel price increases, with adverse consequences for economic growth. Such a setback would in turn hamper the delayed expansion in nuclear power, thereby precipitating a further period of rising energy prices and energy shortage. To avoid such a situation, the expansion of nuclear power at a steady and controlled pace, with well-defined targets in the longer term, must start as soon as possible. Western Europe should coordinate its plans for this expansion, to ensure the best use of skills and resources.

It is worth pointing out that this target implies an increased role for electricity in meeting energy needs, with an increase in the proportion of world useful energy supplied by primary electricity from the present figure of $12\frac{1}{2}\%$ to a little over 20% (using the energy coefficient devised by Adams & Miovic 1968). This should be attainable.

It is particularly important that the countries of Western Europe make the necessary effort to achieve these targets. In 1985, even with North Sea oil and gas, the annual oil imports of the Nine will be over 900 million tonnes at a time of great pressure on supplies from other oil-importing nations. It is clearly desirable that Europe should have alternative sources of supply as soon as possible, and of these alternatives nuclear power is currently the only practical contender. The various energy policies put forward both by individual European countries and by the Community recognize this by envisaging a rapid expansion of nuclear power.

Nuclear power, then, is now poised on the brink of a major expansion. Its capital cost disadvantages are becoming less marked at the same time as its undoubted running cost advantages are increasing. Moreover, a large increase in nuclear power is necessary if an 'energy shortfall' in 20–30 years time is to be avoided, and this country now has a nuclear industry capable of meeting the challenge of such an increase. Finally, we must not forget the advantages of an increased nuclear power programme for the balance-of-payments situation and in economizing altogether in the use of oil.

At present, nuclear plants account for about 7% of this country's electricity generating capacity, and by 1977/8 when the a.g.r. have all come into operation this proportion will rise to about 15%. Over 10% of Britain's electricity is currently generated by the nuclear stations, and this proportion is expected to rise to over 20%. By 1978 our power-stations will be saving the equivalent of 15×10^6 tonnes of oil a year (17.5×10^6 m^3), and even a modest nuclear programme would double the nuclear capacity by 1985 to give a saving of 30×10^6 tonnes of oil a year (35×10^6 m^3). On a conservative estimate of an increase in oil prices to £15/t in 1985, this is unlikely to be worth less than £450 million a year. The comparable saving in oil from a European Community nuclear programme of even 140 GW by 1985, which is very much lower than some of the most recent official targets of desirable capacity, would be of the order of some 210×10^6 tonnes of oil a year (250×10^6 m^3). The offsetting cost of uranium is relatively small, certainly not more than an eighth of the cost of the oil required for an equivalent programme, and expenditure on separative work should by then predominantly be at 'home'.

Can nuclear power fulfil the role I have described, or are there constraints on its development? In the short term there are no resource restraints; the existing problems are institutional or political. The nuclear industries of Western Europe are able, indeed keen, to tackle increased programmes of work.

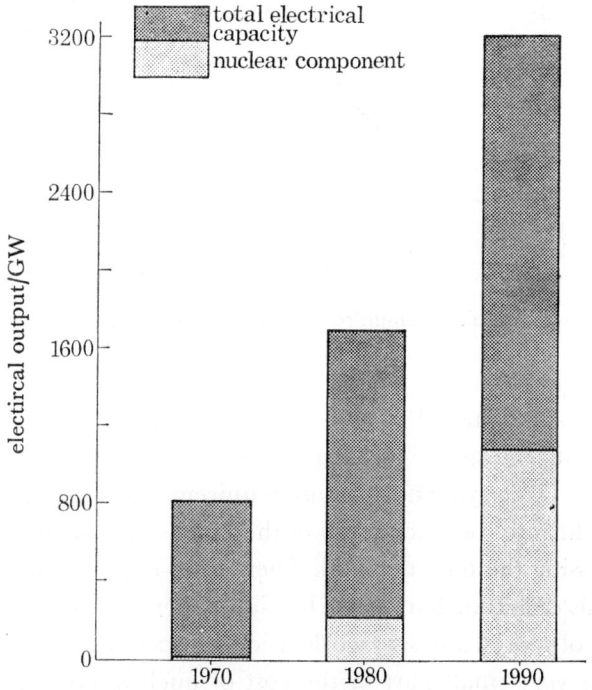

FIGURE 1. World nuclear capacity relative to total electrical capacity.

The thermal reactors already in use and under development could be used to meet the capacity targets called for and are a proved, safe and efficient way of producing energy. But in the longer term, if only thermal reactors are used, problems can be foreseen in providing sufficient reasonably priced uranium to fuel them. This problem has to be seen in a wider context than just Western Europe. A forecast of the growth of nuclear capacity, relative to total electrical capacity, in the non-communist world is given in figure 1 (O.E.C.D. 1973). This forecast assumes that while the rising cost of fossil fuels and their future supply position will make nuclear plant steadily more attractive, the shift in relative economics is not likely to be so radical that all existing coal- and oil-fired stations will be shut down before the end of their working lives. However, the trends in fossil-fuel prices will be sufficient to induce utilities to meet their growing demands for electricity from nuclear plant as far as possible, with fossil-fuel plant being used predominantly as low-merit order and peaking plant to provide capacity at times when system demand is high. By 1990 up to three-quarters of total new capacity ordered each year will be nuclear, with the remainder being divided between the most efficient fossil-fuelled plant, hydro-electricity and specialized peaking plant such as gas turbines.

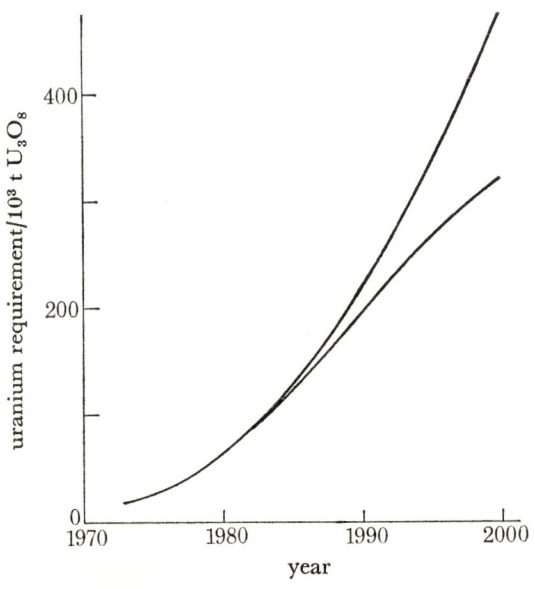

FIGURE 2. Annual uranium requirements for O.E.C.D. area.

If the nuclear capacity was installed at the rate shown in figure 1, and it was all of the thermal type, then the annual uranium fuel requirements would be as shown by the top line in figure 2 (based on O.E.C.D. 1973). The uranium requirements of the various thermal reactor systems do not differ significantly between types of the same size, and the precise mix is therefore not important in assessing the total demand. There is no reason to believe that insufficient uranium exists to meet this demand. But, as Dr Bowie's paper suggests, it may increasingly be available only as a result of more extensive world-wide prospecting and at higher prices. The cost of uranium is only a very small part of the cost of nuclear power, and would increase several times without destroying nuclear power's competitiveness for base-load generation. But the cost of uranium must rise as the lowest-cost resources are exhausted, and there is therefore an obvious incentive to economize in its use if possible.

Such economy will be possible with increasing utilization of the fast breeder reactor, which offers the solution to the world's fuel supply problems for centuries, even if no other solutions are found. The most exciting trend in nuclear power generation in the 1980s will be the increasing introduction of fast breeder reactors into commercial use.

A fast breeder reactor is designed to operate entirely with fast neutrons (i.e. with no moderator), thereby making it possible to convert more than one atom of non-fissile uranium-238 into fissile plutonium-239 for every atom of fuel which is fissioned. To improve the probability of capture of neutrons by ^{238}U atoms, and hence the amount of plutonium produced, a 'blanket' of fertile material surrounds the core of mixed plutonium and uranium fuel. Waste uranium, depleted of its ^{235}U content, is used as the 'blanket' material. Thus, in addition to producing electricity, the fast reactor breeds more new fissile material than it consumes in maintaining the fission chain and this bred material can be used to initiate new fast-reactor stations.

Thus a fast-reactor system utilizes uranium much more efficiently than a thermal-reactor system; typically 60–70% of the uranium is utilized in a succession of irradiations and re-coveries compared with the 1–2% which can be attained with thermal reactor systems. The fuel cycle costs of the fast reactor are, as a result, even more insensitive to variations in the price of uranium. For example, an increase of \$44/kg in the cost of uranium ore, would result in an increase of less than 0.1% of the total fast-reactor generating cost. The world's resources of uranium will be used more efficiently when fully developed fast reactors are installed on a substantial scale and more expensive uranium resources can be exploited thereby ensuring a continuing supply of cheap electricity.

The potential importance of this economy in the use of uranium was soon realized in this country, and led to feasibility studies in the early fifties. It was then clear that the system would require careful and cautious development, but the large potential benefits seemed sufficiently assured for work to start on the construction of the Dounreay fast reactor, completed in 1959. Major fast-reactor programmes have also been started in several foreign countries, most notably France, Russia, the United States, Germany and Japan. Work has continued in this country and a 250 MW (e) prototype fast reactor is being commissioned at Dounreay, which will start delivering power to the grid shortly. Similar-sized prototypes in France and in Russia are at about the same stage of development.

These three countries plan to have their first commercial fast breeder reactors in operation by the early 1980s and to install them steadily from the mid-80s. The other countries too, plan to catch up by then. By the end of the 1980s appreciable numbers of fast breeder reactors will be installed. Figure 3 (based on O.E.C.D. 1973) shows a prediction of the rate of installation of fast breeder reactors in relation to the total build-up of nuclear capacity already shown. Such a rate of fast breeder installation would modify the uranium demand as shown by the lower line in figure 2. The difference between the requirements of the two programmes increases with time as more breeders come into operation, as shown by the illustrative extrapolation of the two lines to the year 2000. The assurance of long-term supplies given by the breeder reactors removes the worry which would otherside exist if only thermal reactors were available. Moreover, this fuel economy is reflected in large savings in the fuel cycle costs. When fast-reactor fuel-cycle costs are compared with light-water reactor fuel-cycle costs, the early commercial fast reactors should have savings equivalent to over £15/kW over the life of the reactor (discounted at 10%) (Marsham 1973). Later fast reactors will use improved fuels, while the fuel-cycle costs for thermal reactors will probably rise; as a result the saving from fast-reactor fuel-cycle costs, relative to thermal, will increase to over £20/kW.

As far as they can be assessed at present, capital costs will be only marginally higher for a fully developed fast breeder reactor than for a comparable thermal reactor. The higher capital costs would offset part of the benefit from the reduced fuel-cycle costs, but nevertheless there should still be a net discounted saving, on a single-reactor basis, of some £10/kW, and rather more if the effect on system costs of the extra plutonium produced is included. As further improvements are made, the savings should increase. Of course, the corollary of the double promise of fast breeders – that they will ensure adequate supplies of fuel *and* reduce the costs of the fuel cycle, is that the longer the delay in introducing the fast breeder, the more concern there must be over fuel supplies and the price of the fuel cycle, as more expensive uranium resources are tapped. The fast breeder is a sophisticated concept, and some degree of caution is essential in the rate at which it is developed. But there is now great confidence in it all over the world, following a long period of painstaking experimentation and development. The first fully commercial reactors of this type will be in operation by the early 1980s. When these first reactors have demonstrated their reliability and economy they will be joined by others as quickly as is practicable, to meet the power needs of the last decades of the century. Design, testing and planning should continue without delay, so that potential problems will have been identified and solved before they can inflict serious or expensive delays.

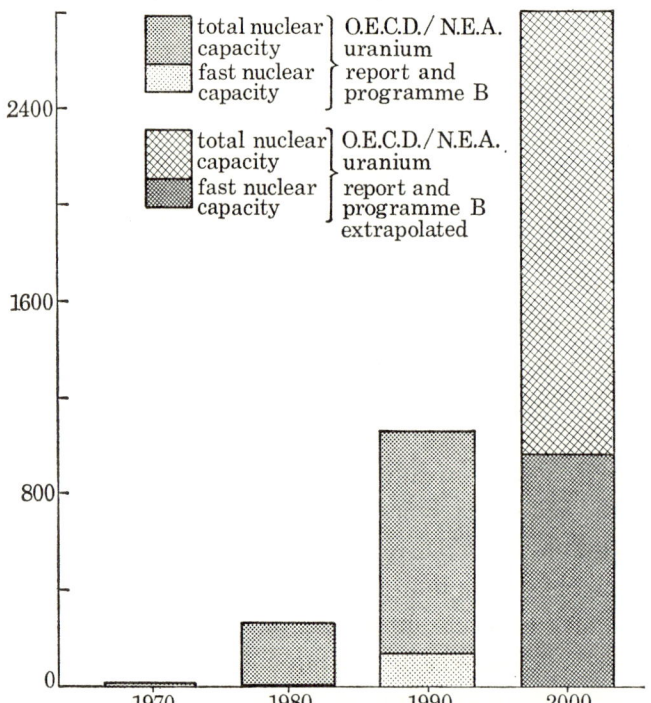

FIGURE 3. Fast-reactor capacity relative to total world nuclear capacity
(excluding communist countries).

This is why the leading industrial nations are now intensively developing the fast breeder reactor. So far these developments have been largely on national lines, but there have recently been moves towards international collaboration. The Germans are collaborating with the Belgians and the Dutch in the construction at Kalkar of a 300 MW (e) prototype fast reactor. In the longer term the French plan to build the successor to their prototype in collaboration with Germany and Italy, this proposed reactor being a fully sized commercial breeder of 1250 MW (e).

This sort of international collaboration is a pointer to the future. Only collaboration in assessing the performance of the various prototype reactors will ensure that the maximum benefit is obtained from those reactors. There must be full collaboration in safety programmes and in formulating commonly accepted safety standards. Furthermore, international collaboration is desirable to ensure the most efficient and economical manufacture of components and fuel and to avoid an uneconomic proliferation of under-used facilities. This country must be ready to play its part in such collaborative programmes especially with her Western European neighbours.

The question of safety in nuclear installations is always uppermost in the public mind and I hardly need add is the first consideration of the designer, builder and operator. Most countries with significant nuclear programmes now have some form of independent Nuclear Inspector, as this country has, whose responsibility it is to inspect, licence and authorize the operation of nuclear plants and to advise Government on nuclear safety matters.

But nuclear safety is international and it is most desirable that standards be accepted internationally. Much good work has been done in the field of radiation protection by the setting of standards by the I.C.R.P. (International Commission on Radiological Protection). In the field of reactor design this international approval has been extended within the European Community by the setting up of a Fast-Reactor Working Group in which all the major nuclear interests are represented. This Working Group has made significant progress in establishing a common understanding of matters relating to safety, and in ensuring that the efforts in all E.E.C. countries make an effective contribution to the E.E.C. fast reactor safety programme.

No attempt will be made here to go into the detail of reactor safety philosophy. However, many ill-informed comments have been made recently in relation to the hazards of 'plutonium reactors' and the transferral of plutonium fuel and a short discussion of this subject seems appropriate.

Plutonium is produced in all existing types of nuclear reactors; fast breeder reactors produce no more plutonium than many thermal reactors per unit of electricity generated and in any case its separation for military use is still a formidable industrial undertaking. The complete elimination of risk of diversion would require the world to abandon nuclear fission altogether as a source of electrical power and to depend entirely on the diminishing fossil-fuel reserves for an unpredictable period, certainly the rest of this century, until alternative sources of energy could be commercially developed.

The extraction of plutonium from irradiated fuel, whether from fast or thermal reactors, and its conversion into a form in which it could be used to make a bomb, requires complex and expensive plant and considerable knowledge. International safeguards are applied to ensure that countries other than those already possessing nuclear weapons do not divert plutonium or other fissile materials from civil to military use and governments impose tight security arrangements for such materials. As regards the physical protection of fissile materials, arrangements are kept under constant review and strengthened as necessary in the light of changing circumstances.

Concern has also been expressed about the problem of disposing of the waste materials produced by the nuclear fission process. The extent of this problem is frequently exaggerated. To a first approximation, the amount of such waste is the same whatever the type of fission reactor used. Such differences as do exist between waste products from fast and thermal reactors do not change the nature of the waste management. Low active liquid waste can safely be released to the sea. Levels of activity are carefully controlled within limits authorized (and monitored) by the appropriate government departments. Small amounts of solids of comparatively low activity accumulate over years. Most of it is stored for long periods in drums in specially provided silos. Its bulk can be reduced by incineration, the ash being retained in the silos; or it can be disposed of in concrete drums in the ocean depths. Highly active fission products are currently stored in liquid form in specially designed tanks and these tanks take up only a small part of each processing site. The safety of the storage (which must be for hundreds of years) would be improved if the fission products were in solid form. Processes for doing this are being developed for introduction when required.

TABLE 2. CURRENT RADIATION DOSES TO U.K. POPULATION

	mrad/year
natural background	100
medical exposure	20
fall-out	2
occupational and miscellaneous	1.5
nuclear industry (radioactive waste disposal)	0.012

Finally, to place the levels of radiation from nuclear plant in perspective, table 2 shows that the U.K. population receives from nuclear waste an annual dose of the order of one-hundredth of 1% of that which is present naturally. Moreover, this proportion will not change significantly even with a substantial increase in nuclear power. Therefore, although there must always be great care involved in the operation of the nuclear-power industry, with rigid observance of the high standards set by international and national safety regulations, I believe that in matters of the environment and safety there are sufficient factual grounds to allow the expansion of nuclear power which, as I have tried to show, is so desirable on other grounds.

This paper has been predominantly about the ability of existing nuclear technology to make a growing contribution to meeting the increasing energy demands of Western Europe in the 1980s in the most economic way possible. It has also indicated the need to plan ahead during this period if longer-term energy needs are to be met. In this connexion it is appropriate to conclude with a few words about fusion power. For the more distant future, fusion power offers the promise of almost unlimited energy with very little in the way of environmental disadvantages. Because of the complexity of the technology, development of controlled fusion has proved inevitably to be a lengthy process, but recently progress has become increasingly encouraging. Fusion scientists in Britain and other parts of the world are now reporting steady progress towards achieving the minimum values of key parameters necessary to achieve a sustained thermo-nuclear reaction. The prize is a great one. Even if we confine ourselves to considering what is believed to be the easiest reaction to achieve (the deuterium/tritium reaction with heavy water and lithium as, effectively, the input fuels) the new source of energy opened up would last the world for thousands of years at a rate of electricity production ten times the present world rate. If deuterium only is needed, the world's seas offer virtually limitless supplies. Most attention so far has been given to fusion involving the magnetic containment of plasma; recently, work has suggested the prospect of controllable fusion using lasers. It is too soon to be certain of the ultimate potential of this method, but it is sufficiently promising to justify continuing study and there is already commercial interest in this approach. A further advantage of a fusion reactor is that there would be few of the environmental problems of radioactive waste disposal; although the central structure of the reactor itself would present problems of disposal at the end of its life its treatment would be similar to that of fission reactors.

It is very hard to say in the present state of knowledge just how fusion power would compare with fission power in terms of the cost of the electricity produced. Fuel costs would almost certainly be an order of magnitude lower even than the very low fuel costs of the fast fission reactor. Perhaps the best way of putting this difference in perspective is to say that it permits the capital cost of the reactor to be about £10 to £12/kW higher than that of a fission reactor while still breaking even with it on total electricity production costs.

In fusion research, however, large experiments may be beyond the existing capacity of any one nation to build, and accordingly the United Kingdom Atomic Energy Authority have recently signed a contract with the European Atomic Energy Community, Euratom, to advance collaboration in research into controlled nuclear fusion and plasma physics. This contract integrates the Authority's fusion programme into an existing coordinated Community programme the aim of which is to construct large experimental installations and, subsequently, prototype fusion reactors for the generation of electrical power. One proposal examined during recent Euratom discussions on future programmes has been the setting-up of a large joint European experiment, the Joint European Tokamak (J.E.T.), to try to establish conditions close to those needed in a power-producing thermo-nuclear reactor. Culham Laboratory will act as host to the Community team which is carrying out the initial design of this experiment. Through these arrangements, the Authority are joining in the furtherance of fusion research in the European Community on a scale comparable with fusion programmes undertaken anywhere else in the world.

Fusion power, when fully developed, will supplement fission power to a growing extent in meeting the energy needs of future generations. But between now and the end of the century it is to fission that Western Europe can turn with confidence to ensure that an increasingly important part of her power requirements is met in a safe, clean and secure manner.

REFERENCES (Hill)

Adams, F. G. & Miovic, P. 1968 On relative fuel efficiency and the output elasticity of energy consumption in Western Europe. *J. Industrial Economics*, **16**, 41–56.
Brookes, L. G. 1973 Towards the all-electric economy. *Atom* **202**, 174–183.
E.E.C. Commission 1973 Guidelines and priority actions under the Community Energy policy. *Bull. European Communities Suppl.* 6/73.
Marsham, T. N. 1973 A U.K. view on fast reactors. *Atom* **201**, 150–163.
Nicholson, R. L. R. 1973 The nuclear power paradox in the U.K. *Energy Policy* **1**, 38–46.
O.E.C.D. (N.E.A.) and I.A.E.A. 1973 Uranium resources, production and demand.

Discussion

A. B. LOVINS (c/o *Friends of the Earth*, 9 *Poland Street, London, W*.1)

I agree that rapid proliferation of thermal reactors would lead to a shortage of low-cost uranium in, as you say, a time of the order of decades. But is it not right that because the cost of uranium is such a small fraction of the sent-out cost of nuclear electricity, the cost of uranium could rise by an order of magnitude (giving us several centuries' supply) without raising the sent-out cost of nuclear electricity by more than a few tens of per cent?

SIR JOHN HILL

The price of uranium could indeed rise by an order of magnitude without significantly affecting the cost of nuclear electricity provided that fast reactors were to be installed in substantial quantities by the time that the envisaged uranium price increases were effective. In thermal reactors an order of magnitude increase in the price of uranium would about double their generating cost.

K. E. ZIMEN (*Hahn-Meitner Institute of Nuclear Research, Berlin (West)*)

The limits to fission energy production will probably be set by the environmental effects of the extremely long-lived isotopes of certain transuranium elements (Np, Pu, Cm) in the radio-active waste from reprocessing plants. Do you agree, or do you think there will be a practicable solution to this problem in the year 2000?

SIR JOHN HILL

It is recognized that it will be necessary to solidify high-activity liquid wastes (which will contain the large majority of the trans-uranic elements) in order to make their containment less dependent on human surveillance. Development of a suitable solidification process has been in hand for some time and it is planned to begin waste solidification in the mid-1980s by means of vitrification. It is, therefore, reasonable to say that by the year 2000 there will indeed be a practicable solution to this problem, and future generations will not find it unmanageable.

K. E. ZIMEN. In my opinion we cannot leave this problem to future generations for two reasons. (1) As a chemist working in reactor chemistry I am convinced that it will be practically impossible to meet the extreme requirements for keeping, for example, Pu away from the biosphere. (2) If this is right and fission energy cannot be used for more than – say – 50 years, then we have to ask: do we really need breeder reactors? For 50 years or so we have plenty of uranium. – I apologize for going beyond the 1980s, but we need to foresee what could happen to the human environment in the next century to be able to make appropriate decisions in the near future.

SIR JOHN HILL

Discharges of radioactive wastes to the environment are regulated by law. These regulations apply of course to plutonium and therefore any discharges of this substance are, and will continue to be, carefully regulated and the standards regularly reviewed.

It is only true to say that we have sufficient uranium for the next 50 years if the breeder reactor is introduced on a substantial scale. Already the world-wide programme of thermal reactors is such that those installed by 1981/82 will require all the known resources of cheap uranium to ensure their fuel supplies for the first twenty years of their lifetime.

PROFESSOR EDWARD EISNER (*Department of Applied Physics, University of Strathclyde, Glasgow*)
Safety in a power system based on nuclear fission: sabotage

It seems to me that the safety of nuclear power systems is usually discussed in terms of accidental failures, and the standards achieved certainly look very impressive. However, I suggest that, in the modern world, an enormously greater risk is that of sabotage. Here a nuclear-fission system differs from a fossil-fuel system in that the sabotage not merely disrupts the system, but could cause widespread danger by release of radioactive material. Could Sir John please give us his thoughts on this?

SIR JOHN HILL

The possibility of sabotage, leading to the release of radioactive material, is, of course, recognized and security arrangements for the protection of nuclear materials have been drawn up in consultation with all interested Government departments. These arrangements are constantaly reviewed to counter this threat as adequately as possible.

Phil. Trans. R. Soc. Lond. A. **276**, 603–609 (1974)

Printed in Great Britain

General discussion

PROFESSOR E. EISNER (*Department of Applied Physics, University of Strathclyde, Glasgow*)

The demand for energy

Throughout this discussion we have heard about the many sophisticated considerations involved in the management of the supply, production, and conversion of energy to meet the estimated future demands. But I have been very disappointed that we have had no paper that discusses the assumptions made in this projection. As far as I can see, such projections are more-or-less sophisticated extrapolations of past trends. Thus, it is assumed that our society will be free to make whatever demand it likes, and that industry must begin to meet this demand. However, I suggest that our society is not a baby that must have demand feeding.

The need for energy depends on per capita production, on energy use per unit of production, and on population. All these have histories, mechanisms, and methods of management, and discussion of these underlies everything we have talked about. Indeed, I suggest that without such a discussion all the rest has been unrealistic.

We seem to be in danger of making the assumption that demand cannot be managed, and that we therefore have to plan to devote huge resources to supplying energy at a rate that is perhaps quite unnecessary. If we devote our resources to this, there will be other things that will go short, many of them things that are vital to a civilized society.

The trouble is that, if our energy producers and converters get all the capital they are asking for, governments are subsequently going to have to show that all this capital really had to be spent, and they will therefore never have an incentive for proper management of the demand. I suggest that we must start by trying to find out how we can reduce the demand for energy made by our society to the lowest rate that is compatible with our goals, and then to ensure that the energy producers and converters do not have any more capital than the least that can possibly meet that demand. Only in that way will there be any pressure to see that our resources are used efficiently.

DR A. S. KENKARE (*Hatfield Polytechnic, Hatfield, Herts*)

Two-tier energy supply system

It is very astonishing that in spite of the looming dangers ahead, on the energy front, there is a strong element of competitiveness in the major suppliers of energy such as coal, oil, gas and electricity. I find this most disconcerting, and only hope that this is a passing phase – the last flicker of the competitive flame, subject no doubt to the formation of an energy board, which seems to be the fashionable idea of the moment. I also hope that the proposed energy board rationalizes the use of energy which will lead to its optimum use.

What is also required in addition is the widest diversification of the energy supply base, so as to include other forms of energy such as solar, wind, tidal, geothermal and energy from waste both physical and biological. Although these 'other' forms of energy add up to a small amount in energy charts at present, their continued use and development could increase this proportion, and what is even more important, make people more receptive to using what in effect is a renewable source of energy supply. The present major energy suppliers almost constitute a cartel and individual consumers have no option but to draw energy from them. By giving encouragement to non-conventional but renewable energy usage, within a framework of safety, people would be able to draw up their own energy schemes with these resources, at least in housing and the non-industrial sector. The distribution of energy supplies by grids is very vulnerable to pressure on fuel resources, and the decentralized arrangement of building 'autonomous houses' that draw minimal amounts of energy from external sources could at least ensure our physical survival in times of scarcity.

In effect, I would like to advocate a two-tier energy-supply system, the industrial and commercial usage of energy being ensured by the four major suppliers, assuming that these are available, and the non-industrial, non-commercial usage having the option, or rather the encouragement, to utilize non-conventional sources of energy supply. It is by these latter options alone that any further development and innovation in energy utilization of non-conventional sources can be made. To me at least, it appears that the near monopoly of the major energy suppliers is seriously inhibiting the innovative potential in so far as the development of non-conventional energy sources is concerned, economic considerations notwithstanding. The so-called energy crisis is really a demand–supply–management problem, and the sooner we adjust the mix the easier will be the problem in the 1980s and beyond.

P. L. Harrison (*Central Electricity Research Laboratories, Kelvin Avenue, Leatherhead, Surrey*)
Solar power

It has been suggested that in the United States of America 3% of the energy required for space heating and cooling could be supplied by solar energy in the late 1980s. If that situation obtained in this country also, what effect would this effective reduction in demand for energy from conventional sources have on the projections that have been discussed at this meeting?

I would draw attention, again, to two factors relating to the use of solar energy as a primary energy source which I think are of particular importance. The first of these is the lack of thermal pollution associated with its use. The second is that, unlike those energy resources that rely on the exploitation of Earth-bound minerals, the cost per unit of energy does not increase with continued use of the solar energy.

Dr J. C. McVeigh (*Brighton Polytechnic and International Solar Energy Society, U.K.*)

One of the questions which should now be receiving serious consideration in the United Kingdom is the extent to which the present knowledge of solar-energy utilization could be implemented in the 1980s. Countries which have been mentioned as having significant nuclear research programmes – U.S.A., France, the U.S.S.R. and Japan – have various solar energy research programmes and many applications. For example, it is estimated that there are well over 2 million domestic solar water heaters in Japan.

Over 20 years ago the late Professor Harold Heywood (*Engineering* **176**, 377, 409 (1953)) first showed that it was feasible to collect an appreciable proportion of the domestic hot-water requirements of an average house in the United Kingdom by using a simple flat-plate solar collector.

It is perhaps surprising to appreciate that even in mid-November it is possible to collect over 4 MJ/m² by using a suitably inclined flat-plate collector. Typically, water temperatures in the order of 40 °C can be obtained from inlet conditions of 10 °C. During the summer period the amount of heat which can be collected is over 10 MJ/m² and water temperatures greater than 60 °C can be achieved.

There are many other applications of basic and applied solar-energy research which could be developed in the United Kingdom and it is very encouraging to note that the United Kingdom Branch of the International Solar Energy Society has been formed with the active cooperation of the Royal Institution.

Dr I. Fells (*Department of Chemical Engineering, University of Newcastle upon Tyne NE1 7RU*)

Manpower

I want to bring to your attention a resource which has not yet been mentioned today but which may have a limiting effect on many of the proposals put forward by Mr Walker and Mr Grainger, that resource being trained manpower. Anyone from the chemical engineering contracting industry must have been delighted by the many suggestions made today to build gasification plant, synthetic oil plant and so on. I am sure contractors if asked if they can cope with such a massive construction programme will say 'yes' because they always do. But they may have difficulty in obtaining the trained chemical engineers to implement the programme. The number of students studying chemical engineering has fallen by almost one third during the last five years. The lead time in education is a long one, schoolboys and girls have to be convinced that chemical engineering offers a rewarding career and it then takes four or five years before they emerge from the universities as qualified chemical engineers.

In some ways it is even more alarming to realize that if a programme of energy conservation is to be implemented, the number of qualified fuel technologists required cannot possibly be met from the 30 or so coming on to the market each year.

This shortage of trained manpower could well limit the programmes necessary to provide oil from indigenous sources in Britain and the U.S.A. to fill the projected shortfall in oil imports.

Professor J. M. Cassels

On the question of manpower, I would like to point out that no immediate action by Whitehall is required. The universities already have the physical resources for a considerable expansion, and they also have in post some of the teachers that would be required. In practice, however, the total university student population is static at present, and within this total the scientific and technological components are declining. The reason is simple: the low level of industrial recruiting among the graduates of one, two and three years ago has sent a wave of dismay through the youth of the country. More vigorous recruiting could be expected to change the situation quite rapidly.

Professor F. J. Weinberg (*Department of Chemical Engineering and Chemical Technology Imperial College, London, S.W.7*)

Combustion of waste

We are doing some research into the combustion of very poor fuels and very lean fuel/air mixtures, some of which are in fact well below the normal flammability limits. This is potentially applicable to burning upcast gas from coal mines and coal seams, exhaust gases from a wide variety of industrial processes, lean methane/air mixtures, wastes fermenting in air, dried sewage, etc. The methods used are based on extensive heat recirculation and on the injection of free radicals into the combustion zone using small plasma jets.

It has been calculated that the entire power requirement of the coal industry in the U.K. could be derived from the methane content of the ventilation air from gassy mines. Even dried excreta contain half the energy value of coal. These are obviously not prime fuels, suitable for every application, but some of them are continuously renewable, some contribute to pollution if they are not burned, and every addition to available energy is obviously to be welcomed. It is likely that in future we shall have to use more and more degraded materials as fuels.

What seems to be lacking from the very interesting compilations presented at this meeting is any estimate of the potential contributions from such sources. Taking the very simplest, what is the total energy value of all the refuse produced by a highly industrialized society?

Mr T. M. Fry (*Tanners, Northbrook, Micheldever, Near Winchester, Hants*)
Energy content of world's reserves of different fuels

At the suggestion of Professor Kurti, I would like to contribute some figures concerning the relative magnitudes of the world's reserves of different types of fuel, expressed in terms of energy content (see table 1). The figures have been compiled in haste from sources that happened to be at hand and are intended to do no more than indicate orders of magnitude, though they are given to one significant figure. In preparing this contribution for inclusion in the published proceedings, I have made some revisions in the data, amplified the commentary upon them and added references, but the table should not be regarded as definitive or even as representative of the best available estimates.

TABLE 1. ENERGY CONTENTS OF THE WORLD'S INITIAL RESERVES OF FUEL

fuel	quantity recoverable	energy content 10^{21} J
coal and lignite	7.6×10^{15} kg	200
oil		
petroleum liquids	2.7×10^{14} kg	10
tar-sand oil	4.1×10^{13} kg	2
shale oil, minimum	2.6×10^{13} kg	1
shale oil, maximum	$(4.5 \times 10^{14}$ kg$)$	(20)
natural gas	2.8×10^{14} m³	10
uranium		
thermal reactors, once-through fuel cycle	3.5×10^{9} kg	1
thermal reactors, plutonium recycle	$(6 \times 10^{9}$ kg$)$	(10)
fast breeder reactors	4×10^{10} kg	2000
thorium	?	(100)
lithium		
cheaply exploitable reserves on land 60% utilization	1×10^{11} kg	500
1% utilization of oceanic lithium	2.5×10^{12} kg	200000
fuel consumption to 1990, cumulative		7

Coal and lignite

A figure of 7.6×10^{15} kg for the world's initial reserves of minable coal and lignite is quoted by Hubbert (1971) from estimates by Averitt (1969), both of the United States Geological Survey. It represents 50% of the initial coal in place in beds not less than 0.35 m thick and at depths not greater than 2 km. Although he uses this figure, Hubbert suggests that a lower figure of 4.3×10^{15} kg may be more realistic in view of the thinness and depth of some of the coal seams included in Averitt's estimates.

With regard to distribution, Averitt estimated that 65% of the initial supply occurs in a region comprising Asia and European U.S.S.R., 27% in North America, 5% in Western Europe and only 2.4% in the three entire continents of Africa, South America and Australia.

In converting to energy content, Hubbert assumes a factor of 2.6×10^7 J kg^{-1}.

Oil

Hubbert's estimate of the initial quantity of recoverable petroleum liquids is 2.7×10^{14} kg (32×10^{10} m^3), of which about five-sixths is in the form of crude oil and one-sixth 'natural-gas liquids' (Hubbert 1971).

The figure for tar-sand oil relates to the Alberta deposits only, no world inventory being available to Hubbert (1971).

The figure for shale oil quoted by Hubbert (1971) refers to the quantity regarded as recoverable under conditions prevailing in 1965, i.e. 2.6×10^{13} kg (3×10^{10} m^3). If marginally recoverable material is included, an upper figure of 4.5×10^{14} kg (52×10^{10} m^3) may be considered (B.P. Co. Ltd 1973).

The energy content factors used by Hubbert (1971) were 4.3, 4.4 and 4.6×10^7 J/kg for petroleum liquids, tar-sand oil and shale oil respectively.

Natural gas

Hubbert (1971) based his estimates for natural gas on those for crude oil, on the grounds that, for large productive areas, a roughly constant proportionality exists between them. His figure is 2.8×10^{14} m^3 and his energy content factor 3.75×10^7 J m^{-3}.

Uranium

The magnitude of recoverable uranium reserves is dependent upon the price that they could command, which in turn depends upon the efficiency with which they could be utilized. For currently operating reactor systems, which use natural or enriched uranium on a once-through basis, the efficiency of uranium utilization ranges from 0.2% for Magnox to 0.8% for Candu, according to Moore (1969). With regard to future developments, plutonium recycling in thermal reactors might increase uranium utilization to about 2%, while the alternative of plutonium recycling in fast breeder reactors might result in 60% uranium utilization.

The quantity of uranium likely to be recoverable within a cost limit of the order of \$30/kg (£12.50/kg) is here assumed to be 3.5×10^9 kg. To put the cost in perspective, the contribution that it would make to generating costs, assuming 0.4% uranium utilization and 33% thermal efficiency, would be 11p/GJ (0.04p/kW h). With regard to lower-grade ores, it has been estimated that 6×10^{10} kg are available at costs up to \$200/kg (£83/kg) (Institute of Fuel 1973), but I do not know whether full account has been taken of the possible restrictions that health and environmental factors may impose. Here, it is assumed that 1% of the uranium in the world's oceans could be recovered within a cost limit acceptable for fast breeder systems, yielding a supply of 4×10^{10} kg.

The energy content of natural uranium, on the basis of nuclear fission of both ^{235}U and ^{238}U, is 8.3×10^{13} J kg^{-1}.

Thorium

The total amount of thorium in the lithosphere seems to be three or four times as great as the total amount of uranium, but the amount recoverable at low cost may be less. The efficiency with which it could be utilized is uncertain, but it is possible that a high figure could be achieved by recycling ^{233}U in either thermal or fast reactors.

The energy content of thorium as a fuel for fission reactors is about the same as that of uranium.

The figure shown in the table for the recoverable energy content of the world's thorium reserves cannot even be regarded as indicative of the order of magnitude.

Lithium

Known and inferred reserves of lithium in the United States amounted, in 1970, to about 6×10^9 kg. At that date, known and inferred reserves elsewhere amounted to only 2×10^9 kg, but this figure may merely reflect the lack of incentive for exploration. If, as seems probable, lithium reserves are distributed evenly, world-wide prospecting should reveal reserves of the order of 1×10^{11} kg, exploitable at a cost of \$20/kg, the price level held by lithium metal during the 1950s and 1960s (Holdren 1971). The lithium content of sea-water is 0.17 g m^{-3}. If 1% of this could be recovered at a cost acceptable for fusion reactor systems, the oceans represent a reserve of 2.5×10^{12} kg.

The energy content of lithium as a fuel for the deuterium–tritium fusion reactor would depend on the details of how it was used. Maximizing for lithium utilization by exploiting the endothermic interaction of fast neutrons with ^7Li, the energy content factor for natural lithium may be as high as 9×10^{13} J kg^{-1} (Holdren 1971). It is noteworthy that Hubbert (1971) used a factor of about 3×10^{13} J kg^{-1}, which corresponds to the consumption of the ^6Li isotope only.

Deuterium

The development of the deuterium–deuterium fusion reactor would open the way to the utilization of a much larger energy reserve.

World energy consumption

Assuming a 5% per annum growth rate in world energy consumption throughout the 1970s and 1980s, the total cumulative consumption of world fuel reserves by the end of the 1980s will have amounted to about 7×10^{21} J.

References

Hubbert, M. K. 1971 *Environmental aspects of nuclear power stations.* Paper IAEA-SM-146/1, International Atomic Energy Agency, Vienna.
Averitt, P. 1969 *Bull. U.S. Geol. Surv.* p. 1275.
British Petroleum Company Limited, Policy Planning Staff. 1973 *World Energy Prospects.*
Moore, R. V. 1969 *Atom* **149**, 59–74.
Institute of Fuel Working Party. 1973 *Energy for the future.* London: Institute of Fuel.
Holdren, J. P. 1971 *Adequacy of lithium supplies as a fusion energy source.* UCID-15953 Livermore, California: Lawrence Livermore Laboratory, University of California.

Mr D. C. Ion (*Britannic House, Moor Lane, London E.C. 2*)

The alternative listing of energy statistics by Mr Fry are welcome but the reader should be aware of serious limitations which arise. In brief terms, the kind of problems involved in use of figures of this kind may be listed as follows:

(*a*) I agree with King Hubbert's caution about Paul Aberitt's coal figures but would much prefer the lower figure of 4.3×10^{15} kg. This itself represents a very much higher recovery ratio than is now adopted (for example) by the National Coal Board, and the reserve-concept behind these figures is very different from that of the petroleum figures.

(*b*) Perhaps the 2.7×10^{14} kg is a reasonable figure for initial recoverable oil reserves, but King Hubbert's reasoning with the gas reserves, using the U.S.A. average, is very dubious.

(*c*) My strongest objection is the adoption of the Alberta figures for tar sands as the total world figure. Venezuela has about the same oil-in-place in its tar sands in Orinoco and there are very big oil reservers elsewhere, in Canada, in Madagascar, and many other places. Some figures were given at the 1973 Institute of Petroleum Summer Meeting and these should be published shortly; see papers by both Leslie and Clegg.

(*d*) I hesitate to comment on the figures for nuclear energy but must express considerable reservation at suggested utilization of 1 % of the uranium from the oceans leading to 2000:1 ratio of fast breeder reactor to thermal reactor resource availability.

(*c*) Finally, I would question whether a table including such extremely long-term considerations has detailed relevance to the next thirty years.

Phil. Trans. R. Soc. Lond. A. **276**, 611–615 (1974)
Printed in Great Britain

FUTURE BALANCE OF ENERGY RESOURCES

Looking ahead

By G. B. R. FEILDEN, F.R.S.
British Standards Institution, London

I have the challenging task of summing up in a notional twenty minutes this fascinating two day Discussion Meeting.

First I will say something about the trends which have emerged from the papers and discussion. We began with the Secretary of State, who gave us a thoughtful paper showing a full recognition of the realities which constrain any power and energy policy. He told us that 'all energy predictions are wrong because of the great difficulty in predicting the demands and supplies'. He instanced this by saying that no one foresaw the North Sea development or the rapid progress with the breeder reactor, and went on to remind us that the balance between fuel sources may alter radically. So, from a policy-making angle we must have a flexible programme. He stressed the need to re-vitalize the coal industry, but warned us that new collieries take many years to bring in to production – as for instance, the 8–10 years for full production from the new Selby colliery.

Turning to the problems of the nuclear power programme he suggested that we would be moving to a situation where development was more international than hitherto. Having begun with several nuclear consortia, we now had a single national design and construction firm with the resources needed for operating on a world scale.

After the Secretary of State's paper, Sir Alan Hodgkin, P.R.S., asked about pooling information between countries. Mr Peter Walker said that he had plans for going ahead with joint research projects. Professor Kurti asked about improving insulation in buildings as a means of saving energy. The Secretary of State agreed that this was very worth while but suggested that, like so many other things, it is not possible to introduce an immediate step function change. There simply was not the industrial capacity to make large-scale improvements in thermal insulation in a short time.

Mr Darmstadter gave us an interesting review of the world energy requirements. A consistent package of policies was needed if a stable energy situation was to be achieved, but he expected the U.S. to weather the crisis. I thought at the time that perhaps he was being a little optimistic, and some later speakers felt the same. He also drew attention to shale and reminded us that this was the third time that the shale-oil proposition has come up, but it has always been overtaken by new discoveries of oil or gas reserves. He drew attention to the difficult environmental problems posed by shale oil.

I was interested that neither Mr Darmstadter nor indeed any other speaker in the whole of this meeting mentioned the possibility of saving oil by altering our ideas on transport. Dr McPhail of Canada has just given me a paper where he says that in the case of Canada – and I imagine the situation is very similar in the United States – a very large percentage of oil is used in private automobiles – about 78% for cars, 6% for aircraft, 3% for trains and 3% for inter-city and urban buses. A doctor friend of mine in Chicago has a Cadillac car with an 11 litre engine which never runs at more than a small fraction of its potential power output. He has told me that its fuel consumption is less than 10 miles per gallon ($\sim$ 4 km/l). My own car is large by European standards, having a 3.5 litre engine which develops a maximum of 135 kW. This will propel the car, whose mass is 1.5 tonnes, at nearly 200 km/h – well over 100 miles/h, and far beyond any road speed limit. So I would suggest to our friends from North America that they should move to the greater use of more economical engines as an immediate means of saving a substantial amount of oil. This is one change which could be implemented relatively quickly to reduce the consumption of high-grade fuels.

Next we went on to an entertaining paper by Mr Ion, who gave us much information, presented in a lively way. In the discussion Professor Kurti and Sir Kingsley Dunham brought us back to the question of units, and here we ought to clear things up before we go any farther. The Royal Society has published a very useful booklet called 'Quantities, units and symbols'. My own organization has published British Standard 3763 on 'The international system of units'. These two documents are entirely in agreement and they draw attention to the fact that the unit of energy is the joule. Many speakers referred to the U.S. petroleum barrel. This contains 42 U.S. gallons or 35 Imperial gallons. Taking the energy value for paraffin, one barrel would deliver some 5760 MJ of heat. Sir Kingsley Dunham also mentioned the unit Q, which I first heard referred to by Dr Huomi Bhaba at the World Nuclear Energy Conference in Geneva in 1955. This unit is 10^{18} erg or 10^5 MJ. The Royal Society Secretariat have assured me that when the proceedings of this meeting are recorded in *Philosophical Transactions*, the SI equivalent will be given for all non-SI units, so comparisons will be much easier than with the variety of units used by speakers.

Coming to coal, Mr Armstrong reminded us that this fuel is surprisingly consistent in quality irrespective of its age or where it comes from. After taking us round the resources available in different parts of the world, he summarized the problem by saying that the geology and exploration can be dealt with fairly easily, but the problem is to get the coal out. In the case of oil it might cost £300M to explore a new off-shore field, but for the exploration of a new coal field the cost might be only £2M. He discussed new methods of mining which can operate in a 1:3 or 1:4 gradient, and ended by saying that basically there is no world shortage of coal but the reserves are badly distributed.

In the discussion Professor Thring made an interesting comment about his Telecheric robot. He said that only 1% of the money which had been put into nuclear-energy research and development would have sufficed for the development of robots for the coal-mines. These robots would be able to win the coal from thin seams, vertical seams or seams far underneath the sea. He urged that imagination was needed to find reliable solutions for the mechanical engineering problems involved. Several experts from the fuel side thought it was nothing like as simple as that: it was a question of getting down to proper engineering design, ensuring that the equipment was completely fool-proof and had large margins for safety. I must agree with the view: any equipment of this kind must be designed for the overloads to which it will sooner or later be subjected in service.

Moving on to oil, we had a stimulating presentation by Sir Eric Drake. His message was summed up in the last of his illustrations, which gave a projection of the main sources of energy to the year 1985. He reminded us that oil scarcity is a real possibility after 1978, and that fossil-fuel resources are being used at a rate which will exhaust them in perhaps 100 years, which is very short in geological time.

During the discussion we had a debate between Mr Ion and Mr Darmstadter on the energy forecast for the United States. Mr Shaw suggested that if you spent enough money looking, you are sure to find. Sir Eric replied that the dilemma is that you may find it, but you are not sure you can keep it: even if you spent millions of pounds in exploration the oil may be expropriated.

On gas, Mr Coppack gave us an account of the known reserves, which fall out roughly at one-third in U.S.S.R., one-sixth in North America, another sixth in the Middle East, and the rest elsewhere. He expected that most new discoveries would be made in the Soviet Union. For natural gas there are great possibilities, for example in the Celtic sea, and we may see gas piped to Europe from the Middle East. He also considered the possibility of methanol as a cheaper means of conveying a gas-based fuel over long distances because tankers are less expensive than the insulated ships needed for taking liquefied natural gas. He then suggested that airships may be used for gas transport, and finally he gave us another slide showing his prediction for gas production in the United States, which showed the expected decline from 1973.

Turning to hydro electricity, Mr Vernon said that 65% of the hydro resources were already used in the developed countries but only 10% in planned economies and a modest 3% in developing countries. This defines the scope for development, and he reminded us that, of the developed countries, Canada had the largest potential for further hydro schemes. Pump storage, with its very useful characteristic of being able to take a sudden peak load almost instantaneously, can be installed without spoiling the environment, as he showed in one of his illustrations. Tidal energy was almost a non-starter because of the very large capital investment costs. Until interest rates fell to 3 or 4% p.a. he foresaw little future for it.

Dr Bowie told us about natural sources of nuclear fuels, beginning with granite which contains about $4/10^6$ of uranium. In the discussion Mr Leslie made the point that the cost of ores is only 5% of the cost of a reactor, and claimed that a breeder reactor will use uranium something like 30 times more efficiently than a thermal reactor.

Professor Leardini showed us the interesting development of geothermal power in Italy and elsewhere. I will remember for a long time his picture of a 'blow-out' with steam gushing out over the Italian countryside. He showed us maps indicating where further geothermal sites might be found all over the world, but emphasized that their total contribution to world energy requirements would only be a modest one.

On our second day we considered energy conversion technology, beginning with Mr Grainger's paper on coal. He reminded us that there is plenty of coal in the ground, that coal is very versatile and that the reserves are sensitive to price. In other words, as the price rises it pays to work the more difficult seams. He then went on to talk about the very important new development of fluidized bed combustion combined with a pressurized 'Coalplex' or coal refinery. For power generation he mentioned the steam/gas-turbine cycle, which is a promising development for improving thermal efficiency, and is readily applicable to coal burning. He again emphasized the need for coordination on a national scale in research and development work spreading into the international field. In the discussion Sir Kenneth Hutchinson spoke about the work of the late Dr F. J. Dent, F.R.S., on the synthesis of methane, the slagging gasifier and other developments which still have a great deal to contribute to the energy situation in many countries.

Speaking about non-conventional hydrocarbons, Mr E. B. Walker reminded us of the enormous world appetite for energy, which meant we will need *all* forms of energy. He suggested that the situation was a continuum and I entirely agree. There was no major synthetic-fuel production going on at the moment but the possibility was attractive for the 1980s. He told us about the tar-sands and U.S. oil shales, and then went on to talk about developments based on coal conversion.

Mr Rooke told us about future trends in gas production and transmission, reminding us of the advantages of gas and that the demand far exceeded the supply. It is a premium fuel and burns to carbon dioxide and water vapour. He talked about pipeline techniques and I was patricularly interested to hear his mention of the new composite wound pipe, with which I had something to do in the early days with its inventor Mr Campbell Secord. This is a development which will have a role to play particularly in off-shore oil pipelines, for which its qualities make it specially suitable.

Mr Clark talked about energy conversion to electricity and gave us an insight into many aspects of the C.E.G.B.'s work. He reminded us that in its earlier years, nuclear power had to face falling real prices of fossil fuels, but I am sure all of us are very glad that the programme was allowed to continue. In the 1980 decade he predicted that new plant ordered would probably be roughly 75% nuclear and only 25% fossil-fuel fired. Finally, he made some interesting comments on transmission, showing us maps of the 400 kV 'supergrid' and said that this installation would see us into the 1980s. In the discussion we began by considering what happens to the waste heat from our power stations up and down the country. Mr Clark had to admit that most of it goes up the cooling towers. He did tell us about an interesting development of oysters growing at Stanswood Bay near Fawley Power Station, but this is a very small usage of waste heat. Several people were rather worried about the wastage and suggested we ought to do more to use the large amount of heat that is thrown away from any power station.

Dr Broom gave a paper on actual operating experiences in the nuclear power stations in this country, and I found this a most interesting account. He reminded us that the first nuclear chain reaction was only a little over 30 years ago, and showed us some of the problems of long-term effects on structures in a reactor and the expedients which have been developed for dealing with them.

We then had Sir John Hill telling us of the plans for further nuclear developments in this country. He reminded us that the industry began with five consortia and that the industrial structure was wrong. There is now a single company for construction and a single company for fuel processing. He looked to the fast reactor consuming 70% of its uranium, and reminded us that this makes the use of dilute ores practical politics. His remarks about dealing with fission products were reassuring on environmental grounds. He claimed that if fission products were scaled into glass with a stainless-steel casing round it, a single pond the size of a small field would look after our total storage requirements to the year 2000. Many people have been over-anxious about that important question.

There followed an interesting discussion on the question of terrorist attacks on nuclear stations, which showed some difficult problems and brought us into fundamental questions of our present social organization on which I will not dwell now.

Let me end by saying a few words about my view of the major changes which are to be expected. First, there can be no doubt that liquid and gaseous fossil fuels will become more expensive. Secondly, it is clear that coal is likely to stage at least a partial comeback, and it will probably be used for the synthesis of natural gas and oils as well as its traditional use in power stations. Thirdly, we can expect a steady growth of nuclear power with more economical systems.

These are likely to be our major energy sources in the 1980s, but there is likely to be progress in the use of solar energy, first directly, but I suspect also via biological processes. I am thinking here of two main routes – what has been called sea-farming, and the use of vegetation in fermentation processes. Fermentation occurs naturally in the production of marsh gas (methane) from vegetation in the bottoms of ponds and lakes, but it is now used on a large scale for sewage treatment. Here, a favourable energy balance is obtained because the activated sludge will produce enough methane to run the prime movers for the sewage works and leave some surplus for sale as a fuel. Another possibility is the production of the so-called 'wood alcohol' from cellulose waste.

Next we have got to think about conserving waste heat from all our power stations. I have great sympathy with our friends from the Generating Board, but I am sure that as fuel costs rise we will have to do more with the vast amounts of low-grade heat which at present are thrown away. We will also see the development of a variety of total-energy schemes. Perhaps this is a new subject to some people. In its simplest form you have a prime mover which produces the electric power needed. You then use the waste heat in winter for heating a building or a complex of buildings. In summer you can have a large absorption refrigerator and use the waste heat for air-conditioning and cooling. An increasing number of installations of that sort are being commissioned in the States, usually fairly small ones, but they deal with large shopping patios and blocks of flats giving a very high total-heat usage. The best case I know of is one in a linoleum factory where 91% of the heat in the fuel is used, thanks to the fact that large quantities of warm water are needed in the process.

I shall end with a prescription for a needed development in energy transmission which would greatly benefit the environment. This is for an underground transmission system for high-voltage electricity. With such a transmission the environment would be no more disturbed than it is by oil and gas pipelines. Whether we shall end up with superconductors or improved insulation with conventional conductors will emerge from research now in progress.